Praise for *Engineering Leadership: The Hard Parts*

Engineering Leadership: The Hard Parts

Navigating Chaos to Build Teams That Deliver

Juan Pablo Buriticá and James Turnbull

O'REILLY®

Engineering Leadership: The Hard Parts

by Juan Pablo Buriticá and James Turnbull

Copyright © 2026 Worthwhile Technology LLC and James Turnbull. All rights reserved.

Published by O'Reilly Media, Inc., 141 Stony Circle, Suite 195, Santa Rosa, CA 95401.

O'Reilly books may be purchased for educational, business, or sales promotional use. Online editions are also available for most titles (*http://oreilly.com*). For more information, contact our corporate/institutional sales department: 800-998-9938 or *corporate@oreilly.com*.

Acquisitions Editor: David Michelson	**Indexer:** nSight, Inc.
Development Editor: Shira Evans	**Cover Designer:** Susan Brown
Production Editor: Clare Laylock	**Cover Illustrator:** José Marzan Jr.
Copyeditor: Piper Content Partners	**Interior Designer:** David Futato
Proofreader: Vanessa Moore	**Interior Illustrator:** Kate Dullea

January 2026:   First Edition

Revision History for the First Edition

2026-01-20:   First Release

See *http://oreilly.com/catalog/errata.csp?isbn=9781098175634* for release details.

978-1-098-17563-4

[LSI]

Table of Contents

Preface. xiii

1. Embracing the Unknown. 1
    What Does Chaos Look Like?   1
        Lack of Ownership   2
        Absence of Structure   2
        Outsized Responsibility   3
        Undefined Process   3
        Poor Communication   4
        Blame Culture, Scapegoating, and Recrimination   4
        High Turnover   5
        Lack of Strategy and Planning   5
        Frequent Crisis Mode   5
        No Meaningful Measurement   6
    Chaos or Opportunity?   6
    Well, That Was Fun   7
    This Sounds Scary   8
    What Are We Going to Do to Help?   9
    Conclusion: From Chaos to Clarity   10

2. Understanding Your Role. 13
    What's Your Job, Exactly?   13
        The Five Pillars of Engineering Leadership   15
        Boss, Where's the Roadmap?   15
        Don't Retreat to the Comfortable   16
        Engineering Leadership Is About Being a "Generalist with Range"   16
        Action Beats Inaction Every Time   17
        Role Description: A Do-It-Yourself Blueprint   17

People: It All Starts with Your Team 18
 What's Your Starting Point? 18
 What's Your Mission? 19
 What Tools Do You Need to Get There? 19
Mission: You've Got a Crew; Now You Need a Place to Go 22
 Where to Start 22
 Getting Your Team on Board 23
 Turn the Mission into Action 23
Plan: You Know Where to Go; Now Chart the Course 25
 Build with Your Partners 25
 Know What to Prioritize 25
 Turn Strategy into Action 26
 Adjust, but Stay Focused 26
Process: The System That Keeps Things Moving 28
 Keep It Lean, but Effective 28
 Make Process a Team Effort 29
 Adjust As You Go 29
Product: What Are We Actually Delivering? 31
 Know What Success Looks Like 31
 Balance Quality with Speed 32
 Outcomes Aren't Always Features 32
Conclusion: Pulling It All Together 34

3. Navigating Chaos. 35
Diagnose Context 37
 Why Diagnosis Matters More Than Style 37
 Four Lenses for Reading Context 38
 The Reality of Power and Identity 40
 Rapid Assessment When You're New 41
 When Diagnosis Reveals Hard Truths 42
 Connecting Diagnosis to Action 42
Thrive in Chaos 43
 The Trap of Bureaucratic Control 44
 The Punk Alternative: Motion Creates Clarity 44
 From Energy to Discipline 45
 Open Source: Punk Principles at Scale 45
 What This Means for Seasoned Leaders 46
 The Political Reality 46
 When Momentum Matters Most 46
Focus on Outcomes 47
 The Sophistication Trap 47

	Why Outcome Frameworks Miss the Point	47
	When Outcomes Are Genuinely Unclear	48
	The Political Dimension	48
	Context Shapes Outcome Focus	49
	Failure Modes of Outcome Obsession	49
	What This Looks Like in Practice	50
	Making It Sustainable	51
	Shift Between Roles	51
	The Cost of Leadership Rigidity	51
	Three Essential Modes	52
	The Switching Skill	54
	The Reality of Constraints	54
	Leading Through Others	55
	What This Looks Like in Practice	55
	Conclusion: Pulling It All Together	58

4. Building Cohesive Teams.................................... 59

	It Starts with Safety	59
	Get a Functional Team, Fast	61
	Understand the Work	61
	Renegotiate Commitments	62
	Remove Blockers	63
	Get Capacity Under Control	64
	Understand Skills, Dynamics, and Performance	65
	Capacity Is Not Enough; Build Capabilities	67
	Capabilities as a Team-Level Tool	67
	What Great Teams Do	68
	Growing Without Losing the Plot	71
	From Supporter to Supported	72
	You Make Yourself Replaceable	72
	You Scale the Culture, Too	72
	The Bottom Line	73
	Conclusion: From Chaos to Cohesion	73

5. Setting Direction....................................... 75

	What Causes Drift	76
	The Drift Diagnostic	77
	The Silent Cost of Drift	78
	What Good Direction Looks Like	79
	Traps That Cause Drift	81
	No Clear Inputs	81

Too Many Inputs	81
Hesitant Leadership	82
Reactive Roadmaps and Consensus Traps	82
Be the Lighthouse	83
Creating Alignment	84
Why Alignment Breaks	85
Alignment Is Not Agreement	85
How Leaders Create Alignment	86
Adapting Without Losing the Plot	87
If the Plan Changes, the Story Stays Intact	88
Clarity Is the Work	88
Reflection	90
Conclusion: Pulling It All Together	91

6. Shipping Products and Code in Chaotic Environments. 93

Inspire Confidence	94
Be Execution-Focused	95
Be Lightweight and Adaptable	96
Foster Continuous Learning	98
Focus on Communication and Collaboration	98
Celebrate	99
Other Prioritization Concerns	99
Not All Work Is Created Equal	100
Not All Work Can Use the Same Process	100
Not All Work Is Known at the Start of Prioritization	100
Prioritization Techniques	101
Key Inputs for Decision Making	101
Utilize Established Frameworks	102
Managing Backlogs Effectively	111
A Good Cadence	112
Approaching Estimates Wisely	113
Balancing Different Types of Work	114
Addressing Technical Debt	115
Monitoring Progress and Adjusting Priorities	116
Communicating Priorities Transparently	117
Not Everything Is a Product	118
When Linear Makes Sense	119
Creating Just Enough Structure	120
Working the Plan, Not Serving It	120
Connecting Linear and Iterative Work	121
Conclusion: Driving to Execution	121

7. Budgeting, Costs, and Vendors. 123

Why Budgeting Matters   124
    Strategic Alignment   124
    Cost Management   124
    Risk Mitigation   127
    Performance Monitoring   127
    Resource Optimization   127
    Stakeholder Management   128
    Innovation and Growth   129
    Organizational Learning   129
    Competitive Advantage   129
Understanding Cost Management   129
    Fixed Costs: The Foundation   130
    Variable Costs: The Flex Players   130
    Step Costs: The Level-Up Costs   130
    Who Pays?   131
    The Human Side of Cost Management   131
Creating and Managing a Budget   131
    Starting with History   132
    Categorizing Expenses   132
    Headcount Planning and Costs   133
    Setting Goals That Make Sense   136
    Creating Your Budget Template   137
    Tracking and Managing   138
    Learning from Experience   138
Vendor Management   139
    The Great Build-Versus-Buy Debate   139
    Picking Your Partners   140
    Managing the Relationship   140
    Outsourced People and Teams   141
Ethics   142
    Taking Care of Our People   143
    Being a Good Neighbor   143
    The Global Perspective   143
    Looking to the Future   144
    Making It Real   144
Conclusion: Budget as Strategy   144

8. Technical Principles and Strategy. 147

Why Does Technical Strategy Matter?   148
Building a Foundation   149

You Need More than Technical Widgets   151
Establishing Technical Principles   153
   Example Technical Principles   154
   Crafting Technical Principles Workshop   156
   Living Your Principles   158
Scaling Without Overengineering   158
   Choose Boring Technology   160
Developing Your Technical Strategy   162
   Choosing Granularity   163
   Balancing Flexibility with Direction   164
   Collaborative Development   164
   Elements of a Technical Strategy Document   164
   The Living Document Approach   167
   Making Technical Strategies Real   167
Conclusion: Enabling Execution with Technical Strategy   171

9. Collaborative Technical Practices and Decision Making. **173**
Shared Technical Principles   174
Communication, Collaboration, and Execution   177
   Communicate, Communicate, Communicate   179
   Collaboration   189
   Execution   192
Decision Making in Chaotic Conditions   201
   Psychological Safety   201
   Inclusive and Data-Driven Decision Making   203
   Continuous Improvement   204
Conclusion: Collaboration Is the Glue   205

10. Metrics That Matter for Engineering Teams. **207**
Why Measure Anything at All?   207
Metrics as a Product   208
   Design Intentionally   209
   The Trap of Want Versus Need   209
   Constructing Questions   211
   Choosing the Right Metrics for Your Context   212
   Take Stock of What You Already Measure   214
Velocity: The Most Misused Metric in Software   215
   Why Velocity Goes Wrong   215
   How to Use Velocity Correctly   216
Build Your Metrics with Your Customers in Mind   216
   The Human Side of Metrics   217

        Balancing Quantitative and Qualitative Data   218
        Humans and Metrics   219
        Make Metrics Visible and Accessible   221
     Metrics Need Ongoing Maintenance, Just Like Code   221
     AI Doesn't Change Your Approach to Metrics   223
     Conclusion: Measurement as Illumination   224

11. Fitting It All Together... 227
     The Symphony of Chaos   228
     The Foundation: People and Safety   228
     The Lighthouse: Direction, Not Drift   229
     The Engine: Process and Execution   230
     The Compass: Metrics and Measurement   231
     The Ecosystem: Building Beyond Your Team   232
     The Evolution: From Chaos to Capability   233
     The Reality: It's Harder Than It Looks   234
     The Practice: Making It Real   235
     The Future: Beyond Survival   236
     The Call: Your Turn to Lead   237
     The End, and the Beginning   238

Index... 241

Preface

When we—Juan Pablo and James—first met, it was in a breakfast group for startup engineering leaders. We both worked at startups with varying levels of chaos and dysfunction. What was interesting to both of us were the similarities between our experiences. Despite different organizations, people, and products, we saw that when rapid change and chaotic situations occurred (startup life™), people and organizations responded similarly—with the same, less-than-optimal outcomes. Our wider breakfast group also shared similar experiences. (The group was part networking and part support system.) Fast-forward 10+ years, and we've both been in big and small organizations where we've had to deal with the same chaos and dysfunction. In recent years, we've worked together on several projects and been able to compare notes in greater depth. What continued to intrigue us was that these recurring patterns persisted.

Over those years, we've developed approaches, strategies, and techniques to handle chaotic environments and achieve better outcomes for our teams and organizations. We don't approach these problems identically, and we each have our spheres of strengths and domains that interest us. Hence, a regular part of our interactions involved comparing notes on areas where we were weaker.

In addition, several significant chaotic events have stressed leaders and their teams in recent years—for example, dealing with COVID-19 or managing teams of engineers living in war zones and facing political and environmental upheaval. Those experiences, along with anecdotes we heard from our peers, suggested that dealing with chaos was an industry-wide challenge for engineering leaders. We decided to codify our developed approaches and techniques, first as a training course and now as a book.

While the ideas we share in this book may help you survive engineering leadership in chaotic organizations and times, we don't believe they are a cure-all. Still, we hope they will support you on your leadership journey. This book contains hard-won knowledge gathered through some tough professional challenges that we'd prefer you not to face, or it might at least help lessen the impact of those experiences you cannot avoid.

Who Should Read This Book

As an engineering leader, you're responsible for your team's success and the delivery of complex technical projects. This is challenging even under the best circumstances, but it can feel downright impossible within a chaotic environment.

You may be a new engineering manager, suddenly responsible for a team and a codebase that you're still learning to navigate. You're trying to figure out how to balance your team's needs with the business's demands, all while establishing your own leadership style and earning the trust of your reports.

Or perhaps you're a seasoned director or vice president, but you've landed in an organization that's growing faster than its processes and structure can keep up. You're working long hours to keep things on track, but it feels like you're always firefighting and never have time for strategic planning or team development.

You might be a technical leader—a staff engineer or architect—trying to lead without formal authority. You see the technical challenges and risks that must be addressed, but need help gaining buy-in and alignment from organizational stakeholders.

Regardless of your specific role or context, leading in a chaotic environment is draining. You likely juggle multiple roles and navigate endless obstacles and surprises. It can be frustrating, stressful, and demoralizing.

But here's the thing: as hard as it is, your leadership matters now more than ever. In the face of chaos and uncertainty, your team needs someone to provide direction, clarity, and support. They need someone to advocate for them and clear the path for their work. They need someone to create a pocket of sanity and stability where they can focus and do their best work.

That someone is you. You have the opportunity and responsibility to be the leader that your team needs. It won't be easy, but it is possible—and this book is here to help. By sharing our experiences and lessons, we aim to equip you with the mindsets and techniques necessary to navigate and succeed as a leader under challenging circumstances.

Why You Should Read This Book

When it comes to books on leadership and management, there is no shortage of options. Walk into any bookstore or browse online, and you'll find shelves upon shelves of titles promising to make you a better leader. Many of these books are excellent—we've read lots of them and have learned a great deal from their insights and advice.

However, we've also found that much of this advice implicitly assumes that you operate in a relatively stable and functional environment. The case studies and examples often feature organizations with clear structures, well-defined processes, and ample resources to invest in new initiatives and professional development.

But what if that's not your reality? What if you're trying to lead in an organization that's more chaotic than calm—where roles and responsibilities are unclear, priorities change by the day, and you're lucky to get through a meeting without being pulled into the latest fire drill?

In these situations, much of the standard leadership advice falls flat. It's not that it's bad advice—it's just not feasible to implement when you're barely keeping your head above water. Telling a leader in a chaotic environment to "just" implement a new process or "just" carve out time for strategic planning is like telling a drowning person to "just" swim to shore. It's not helpful when you're consumed by treading water.

That's where this book comes in. We've both been in the thick of chaotic organizations and had to figure out how to lead and ship products despite the disorder. We've had to adapt the best practices and principles of engineering leadership to fit the constraints and challenges of our environments.

In this book, we want to share what we've learned. We offer a framework for making sense of the chaos and finding your footing as a leader. We provide concrete strategies and techniques for common challenges like clarifying priorities, making decisions with limited information, communicating effectively, and caring for your team and yourself.

Importantly, our advice is grounded in the realities of leading in less-than-ideal circumstances. We know you need more time, resources, and authority. We know you're dealing with ambiguity, politics, and competing demands. Our goal is to equip you with tools and mental models that you can adapt and apply in your unique context.

If you're looking for a playbook to transform your organization into a perfectly oiled machine, this isn't it. But if you're looking for practical guidance on navigating the messy, human realities of leading in times of uncertainty and change, keep reading.

In the book, we'll also walk you through the skills and techniques you'll need to be successful in leading through chaos. The book is divided into two halves. The first part focuses on understanding your role, navigating chaos, setting directions and strategy, and building your team's resilience and skills:

- *Chapter 1*: Understanding what chaos and its symptoms look like
- *Chapter 2*: Understanding your role and creating leadership foundations
- *Chapter 3*: Navigating your role in chaotic environments and learning how to deal with the challenges that chaotic environments throw at you

- *Chapter 4*: Building your team into people who thrive in chaos
- *Chapter 5*: Setting direction for your team

The second half of the book focuses on techniques that you can use to deliver software and set your team up for technical success:

- *Chapter 6*: Shipping products and code in chaotic environments
- *Chapter 7*: Budgeting, costs, and dealing with vendors
- *Chapter 8*: Developing technical strategy and principles that are resilient to chaos
- *Chapter 9*: Building collaboration and making technical decisions in a chaotic world
- *Chapter 10*: Metrics that help your team navigate the chaos
- *Chapter 11*: Wrapping it all up—bringing together the principles and techniques we've shared

We believe that chaos doesn't have to be a barrier to effective leadership—it can be an opportunity. By embracing the uncertainty and learning to lead through it, you can emerge as a more resilient, adaptable, and confident leader. And in turn, you can help your team and your organization find stability and success on the other side.

Conventions Used in This Book

The following typographical conventions are used in this book:

Italic
> Indicates new terms, URLs, email addresses, filenames, and file extensions.

`Constant width`
> Used for program listings, as well as within paragraphs to refer to program elements such as variable or function names, databases, data types, environment variables, statements, and keywords.

This element signifies a tip or suggestion.

This element signifies a general note.

O'Reilly Online Learning

O'REILLY® For more than 40 years, *O'Reilly Media* has provided technology and business training, knowledge, and insight to help companies succeed.

Our unique network of experts and innovators share their knowledge and expertise through books, articles, and our online learning platform. O'Reilly's online learning platform gives you on-demand access to live training courses, in-depth learning paths, interactive coding environments, and a vast collection of text and video from O'Reilly and 200+ other publishers. For more information, visit *https://oreilly.com*.

How to Contact Us

Please address comments and questions concerning this book to the publisher:

O'Reilly Media, Inc.
141 Stony Circle, Suite 195
Santa Rosa, CA 95401
800-889-8969 (in the United States or Canada)
707-827-7019 (international or local)
707-829-0104 (fax)
support@oreilly.com
https://oreilly.com/about/contact.html

We have a web page for this book, where we list errata and any additional information. You can access this page at *https://oreil.ly/engineering-leadership*.

For news and information about our books and courses, visit *https://oreilly.com*.

Find us on LinkedIn: *https://linkedin.com/company/oreilly-media*.

Watch us on YouTube: *https://youtube.com/oreillymedia*.

Acknowledgments

Juan Pablo

To all the nerds who have joined my crew over the years, especially those who keep following me around to increasingly chaotic environments—Bob, Brian, Checho, Elizabeth, Ernesto, Gorsuch, Lara, Nico, Pinilla, and Urbano—you make the impossible feel merely difficult.

Camille, thank you for the walks that helped me think through hard problems; Renee, for the perspective that kept me grounded; Kellan, who coached me out of chaos and showed me that leadership is not about having answers but about asking better questions.

The FiasCo and Mimosaconf crews taught me that the best ideas emerge when smart people feel safe enough to disagree.

Ann, you somehow keep hiring me. You're stuck with me.

Special thanks to James (friend, coauthor, and colleague), without whom this book literally would not have happened. Somehow, he still talks to me after we survived writing it together.

A mi mamá, que si no la nombro no me vuelve a hablar.

Finally, Juliana, whose patience with my writing obsessions made everything else possible.

James

Thank you to the wonderful folks I've worked with over the last few years, who helped us all survive the chaos.

Thanks to my amazing CTO dinner group: Camille Fournier, Peter Miron, Harry Heymann, Kellan Elliott-McCrea, and Illia Papas, who have always managed to make the chaotic slightly more palatable or provided wine when that didn't work.

Thanks to Renee Orser and Monique Garth for good-naturedly listening to me rant about my work on numerous occasions.

Finally, thanks to my best friend and partner in life, Ruth, who has again put up with more nights of me tapping at a keyboard—last book, I promise.

CHAPTER 1

Embracing the Unknown

It's essential to know what you're up against when trying to lead in a chaotic environment. Understanding the nature and extent of the chaos is crucial for effective leadership, as it enables you to identify the root causes of dysfunction and address them strategically. This awareness helps you to navigate challenges more effectively, prioritize critical issues, and implement solutions to restore order and drive progress.

What Does Chaos Look Like?

Sometimes, it can be hard to tell whether your organization is chaotic or if there's just some fuzziness around the edges. Chaos can become a creeping normality. There's a fable about boiling frogs: a frog placed in boiling water will immediately jump out, but if placed in tepid water and heated gradually, it won't perceive the danger and will boil to death. That fable, while factually incorrect, has parallels for how chaos can creep up within an organization.

Your ability to assess the level of chaos can also be variable, especially when starting a new role. In these situations, sometimes you know what you're getting yourself into; your new employer has been transparent about their issues. But sometimes the challenges are presented to you in a rosier light than is really the case. Indeed, you've probably interviewed for a new job in a new environment; you put your best foot forward, and the company, in turn, tried to sell you on their mission, product, and culture. But lo and behold, when you start, you discover challenges and chaos existing that you did not expect.

Even if you're already part of an organization, chaos can sometimes be more evident from different perspectives. James once worked in a large corporate environment where several teams were functional because their leaders shielded them from the broader organization's dysfunction. The organization's state was revealed more accurately when James was promoted to leadership level.

Understanding the symptoms and behaviors in chaotic environments can also be helpful, even if you know things are chaotic—especially if you're going to develop strategies to work through, over, around, or under that chaos.

Chaos can manifest in many ways within an organization. Some common symptoms include the following:

- Lack of ownership
- Absence of structure
- Outsized responsibility
- Undefined processes
- Poor communication
- Blame culture, scapegoating, and recrimination
- High turnover
- Lack of strategy and planning
- Frequent crisis mode
- No meaningful measurement

Let's dive deeper into some of these areas.

Lack of Ownership

In a chaotic environment, it is often unclear who is responsible for what. This ambiguity gives rise to a range of problems. Tasks and decisions can easily fall through the cracks because no one is clearly accountable for them. Confusion and delays are common; either people assume someone else is handling a task, or multiple individuals might redundantly attempt to tackle the same issue.

This lack of ownership makes it difficult to resolve issues or implement improvements. Without clear responsibility, it's easy for things to devolve into a blame game or a state of learned helplessness where everyone waits for someone else to step up and take action.

Absence of Structure

Closely related to the lack of ownership is a lack of organizational structure. In a chaotic environment, teams and roles are often poorly defined. Individuals often struggle to understand their place within the broader context and the impact of their work on overall goals.

This lack of structure makes collaboration and decision making extremely difficult. Without clear lanes, teams often duplicate effort or work at cross-purposes. The left hand doesn't know what the right is doing, leading to wasted resources and conflict.

Outsized Responsibility

When ownership and structure are lacking, it's common for individuals to assume outsized responsibility relative to their role or level of experience. A junior engineer might find themselves as the sole owner of a critical system. A manager might be responsible for multiple teams and products without sufficient resources or support.

This situation is a recipe for burnout. Individuals are stretched too thin and lack the experience or authority to handle their responsibilities effectively. It also creates significant risk for the organization, as a single person becomes a bottleneck or single point of failure.

Undefined Process

In chaotic organizations, processes are often made up on the fly or exist only as institutional knowledge. Documentation is lacking, and the way work gets done is inconsistent and unreliable.

These gaps make it extremely difficult to onboard new team members. Without transparent processes to follow, new hires struggle to become productive contributors. It also means that the organization often reinvents the wheel, with each team or individual creating their own way of doing things.

This lack of process creates resistance to change and stifles innovation. People cling to what they know "works," even if it's suboptimal, because change introduces the risk of failure. Turning inward and being insular also breeds a culture of "not invented here," where the organization aggressively rejects external innovation.

We know this sounds contradictory. Indeed, how can a chaotic organization—where nothing is stable and things change minute by minute—also be risk-averse and resistant to innovation? People become accustomed to their situations. The chaos becomes predictable. We've both had situations where people have reached a state of fatalistic acceptance of their organizational chaos—or even worse, they believe any change will only deepen the dysfunction. Attempts to address that chaos must defeat both this fatalism and the underlying causes of the chaos.[1]

Without a clear and safe way to evolve processes, the organization stagnates.

---

1 Albeit, sometimes chaos and fatalism are self-reinforcing: there is chaos, so we become fatalistic, which in turn generates more chaos.

Poor Communication

Communication breakdowns are common in chaotic organizations. Information often becomes siloed within teams and isn't effectively shared. Updates and decisions are communicated haphazardly, if at all, and relevant stakeholders are left out of key conversations.

Sometimes, there is an active effort in chaotic organizations to restrict communication—either because "no news is good news" or because communicating the current state can make you or your team a target of blame or recrimination.

This effort leads to people feeling out of the loop and disconnected from the bigger picture. It breeds confusion, resentment, anger, and a lack of alignment. Like the lack of structure, this also results in duplicated effort and teams working at cross-purposes.

Blame Culture, Scapegoating, and Recrimination

In a chaotic organization, a blame culture often prevails. This manifests as individuals or teams looking for someone to blame when things go wrong rather than focusing on solving the problem. This culture discourages risk-taking and innovation, as employees fear the repercussions of failure.

Blame culture creates an environment of fear where employees are reluctant to take ownership of their work or make decisions. This hesitancy can lead to stagnation and a lack of progress; individuals become more focused on avoiding blame than contributing positively to the organization's goals. Over time, this will significantly erode trust and morale within the team.

James worked in an environment where, before any attempt to resolve an incident, there was a cycle of "whose fault is it?" One of his colleagues coined the term "blamestorm" to describe the incident-management process. Repeatedly, these efforts to assign blame added significant delays to the recovery time. In turn, this led to conflict and friction among internal teams, external vendors, and leadership.

Scapegoating is another symptom in which individuals or teams are unfairly blamed for systemic problems. This can lead to high turnover, low morale, and a toxic work environment in which employees prioritize self-preservation over collaboration and progress.

Scapegoating fosters a hostile work environment where collaboration is stifled and fear prevails. It diverts attention from addressing the root causes of issues and instead creates an atmosphere of mistrust and division. Employees may become disengaged and disconnected from the organization's mission, exacerbating the chaos.

Ultimately, this results in recrimination, characterized by retaliatory accusations and back-and-forth blame exchanges. This hampers effective problem-solving and deteriorates trust among team members and across departments, further entrenching chaos.

Recrimination can create a cycle of negativity that is hard to break. It often leads to a toxic work culture where energy is spent defending oneself rather than on productive activities. This environment can cause significant emotional stress for employees, reducing overall job satisfaction and increasing turnover rates.

High Turnover

High turnover rates are a significant indicator of chaos. Employees leave due to burn-out, dissatisfaction, and a lack of career development opportunities. This constant churn disrupts team cohesion and results in the loss of institutional knowledge.

Frequent turnover can disrupt work continuity and increase recruitment and training costs. In addition, the loss of experienced employees can weaken the organization's ability to maintain quality and consistency in its operations. High turnover also sends a negative signal to remaining employees and potential hires, damaging the organization's reputation.

Lack of Strategy and Planning

Chaotic organizations often operate purely reactively, lacking clear strategy and planning. Teams and individuals frequently lack a clear sense of purpose or prioritization to guide their efforts. They respond to the latest fire drill rather than proactively working toward defined goals.

Without strategic direction, effort is wasted on low-impact or even counterproductive activities. Resources aren't optimally allocated because there is no clear prioritization of initiatives. The organization ends up frantically treading water rather than deliberately swimming toward a destination.

This disconnect between vision and implementation is also the primary driving force behind fissures between leadership and team members outside the engineering team. These fissures result in distrust and conflict between business units. A classic example is interactions between sales and engineering teams over what to sell and what is feasible to sell, with sales teams overselling and engineering teams overcommitting, undercommitting, or underdelivering.

Frequent Crisis Mode

Organizations in chaos often operate in a constant state of emergency. Every issue is treated as a crisis, which prevents strategic planning and leads to burnout. This mode of operation can also create a volatile work environment, where employees are constantly under stress.

Living in a perpetual state of crisis can create a reactive rather than proactive culture. This environment can prevent the organization from focusing on long-term goals and improvements, as immediate issues consume all attention and resources, leaving little room for strategic planning and development. Chronic stress and burnout can lead to high absenteeism and increased turnover.

No Meaningful Measurement

Finally, chaotic organizations often lack meaningful measurement. They don't define or track metrics that provide a clear picture of the health and performance of the product, team, or business.

Without data, it's nearly impossible to spot problems, measure progress, or make informed decisions. Issues fester unnoticed until they blow up into major crises. Improvements or regressions happen without anyone understanding why. The organization is making decisions in the dark.

This lack of measurement makes it difficult to celebrate successes or learn from failures. There's no reliable way to tell whether an initiative achieved its desired impact or to conduct a productive postmortem on an incident.

Chaos or Opportunity?

But there's a flipside: chaos isn't just something to survive—it's where careers are made. Think about the most respected engineering leaders you know. Ask them about their defining moments. They won't tell you about the time they maintained a stable system. They'll tell you about inheriting a disaster, a team in freefall, or a product held together with duct tape. They became great by creating stability where none existed.

Chaos strips away bureaucracy. In established organizations, changing a deployment process requires committees, meetings, and PowerPoint decks that nobody reads. In chaos, you just fix it. If it works, that's great; if not, try something else tomorrow.

This is the paradox: the absence of structure becomes your permission to build. No ownership? Claim whatever you're brave enough to own. No processes? You won't fight six months to change "how we've always done things." No metrics? You define success.

Undefined roles force you to develop range. You're not just a backend engineer or team lead; you're whatever the company needs that day. You learn product management by necessity. You figure out recruitment because someone has to. You understand business metrics because your engineering decisions suddenly have visible business impact.

The lack of hierarchy creates unusual access. In chaos, you might find yourself explaining architecture changes directly to the chief executive officer. In other companies, that conversation filters through seven layers of management. Chaos collapses distance—your ideas reach decision makers immediately.

But the most significant opportunity is this: chaos reveals who can build versus who can only maintain. The engineers who step up—who bring calm to storms, who ship when shipping seems impossible—are the ones who are remembered, promoted, and recruited. Additionally, when you inherit a mess, nobody expects miracles. Every small improvement is celebrated. (Compare that to taking over a high-performing team where the only direction is down.) In chaos, victories compound. Every fixed bug, every successful deployment, and every new hire who shows up will build momentum.

If you can find the eye of the hurricane and operate from there, chaos becomes your advantage. While others are frozen without structure, you're building it. While they complain about the process, you ship solutions. The opportunity isn't despite the dysfunction; it is because of it. Every missing piece is something essential that you can build. Every fire you extinguish increases your value. Every problem nobody else will touch becomes part of your leadership story.

You're not trying to eliminate all chaos; that's impossible. You're creating just enough order for productivity, just enough process to prevent disasters, just enough structure to maintain sanity. Companies that emerge strongest from chaos have leaders who understand this balance. They didn't impose rigid order on fluid situations—they built systems that could bend without breaking.

Yes, chaos is indeed exhausting and stressful. But it's also where reputations are built and leaders are forged. A year of successful leadership in chaos teaches more than five years of steady-state management. Once you see the opportunity hiding inside each problem, you'll understand why some leaders actively seek chaotic environments. It's not because they enjoy the suffering, but because that's where there are opportunities to do fantastic work.

Well, That Was Fun

If any of those symptoms or the opportunities feel familiar, know you're not alone. Chaos is the reality for many organizations, particularly during periods of rapid growth or significant change. These tumultuous times can exacerbate existing issues and create new ones, making it difficult to maintain order and efficiency. However, acknowledging the presence of chaos is the first step toward addressing it.

Chaos in an organization can be overwhelming, leading to frustration, decreased productivity, and high employee turnover. It can feel as if you're constantly dealing with one crisis after another without making any real progress. This can be particularly challenging for leaders expected to steer the organization toward stability and success. The key to managing and overcoming chaos is recognizing these patterns early on and developing effective strategies to address them.

This involves thoroughly assessing the organization's current state, identifying the root causes of chaos, and implementing targeted interventions to address these issues. It requires strong leadership, clear communication, and a commitment to continuous improvement.

This Sounds Scary

We'd be remiss if we didn't highlight that, after describing the many facets of chaotic organizations, sometimes the best solution is not working there at all. Throughout the book, we underscore the critical importance of self-care, maintaining good mental health, and prioritizing your safety and security. Working in and attempting to improve a chaotic organization is incredibly difficult, stressful, and emotionally draining. The constant flux and unpredictability can take a significant toll on your well-being, often leading to burnout and a sense of helplessness. Leadership, in its own right, is an isolating experience. The responsibility of guiding others, making tough decisions, and being the anchor for your team can create a unique form of loneliness. Leadership in a chaotic environment—where you may lack the support of leaders, peers, and mentors—is incredibly isolating. The absence of a reliable support system and the pressure to navigate through persistent instability will likely amplify feelings of isolation.

We've both burnt out in roles where we felt overwhelmed and isolated. Our experiences have taught us that it's vital to recognize when an environment is detrimental to your well-being and to have the courage to step away if necessary. It's important to remember that no job is worth sacrificing your mental health or personal happiness over. Seeking healthier, more supportive work environments can lead to better professional outcomes and a more fulfilling, balanced life. Additionally, you must consider the long-term impact of staying in such an environment. Chronic stress and burnout can have severe consequences, including anxiety, depression, and other health issues.

When evaluating whether to stay or leave, consider your core values and long-term objectives. Ask yourself if the organization's mission aligns with your personal and professional aspirations. Consider the growth opportunities, both personally and professionally. Are there mentors or leaders within the organization who can provide guidance and support? Is there a path forward that feels sustainable and fulfilling? If the answers to these questions are negative, it may be time to reassess your situation.

Finally, in the spirit of self-reflection, ask yourself, "Am I part of the problem too?" In addition to experiencing burnout and isolation, it's easy to get caught up in the chaos and become a leader who enables or excuses it. Sometimes, in the name of defending your team, you may even become a toxic element in the organization. If you think this is you, consider whether you can change, whether your position is tenable if you try to make those changes, and whether it's worth staying to make them at all.

Moreover, building a solid support network outside of work is essential. Friends, family, and professional networks can provide the emotional support and perspective needed during challenging times. Activities that promote well-being—such as exercise, hobbies, and mindfulness practices—can serve as buffers against the stressors of a chaotic work environment.

Although the drive to improve and contribute to an organization is commendable, it's equally important to recognize when the cost to your well-being is too high. Taking care of yourself should always be a priority. Leaving a chaotic organization is not a sign of failure but a step toward a healthier, more fulfilling professional life. Remember, you can choose environments that nurture and support your growth rather than diminish it. Prioritize your mental health, seek supportive work cultures, and don't be afraid to make bold decisions in favor of your well-being.

What Are We Going to Do to Help?

So, if you want to make a change in your organization, how will this book help you? We aim to equip you with principles, techniques, and mental models to help you lead effectively in chaotic environments. This book is designed to be a flexible toolkit that you can adapt to your specific context, recognizing that each organization and situation is unique. We offer a range of approaches that you can tailor to your specific needs, helping you address the unique challenges you face.

Throughout the book, we explore various topics that are crucial for engineering leaders in fast-paced, growing, and chaotic companies. We discuss the importance of embracing the unknown, navigating minimal structure, and understanding the multifaceted roles and responsibilities of an engineering leader. You'll learn core leadership skills, including adaptability, effective communication, business and product acumen, empathy, collaboration, and strategic thinking.

Next, we will explore how to set direction, build cohesive teams, and scale through systems and delegation. You'll learn to establish goals, develop strategies, align company objectives with team goals, build capable teams, celebrate successes, avoid burnout, and leverage external support. We also discuss implementing efficient systems and empowering teams while managing dependencies.

The next chapters tackle the challenges of managing technical and operational chaos. You'll learn various prioritization techniques, how to balance different types of work (features, technical debt, and maintenance), and how to plan and budget effectively, including vendor management and procurement. We also focus on technical strategy and execution, covering topics such as aligning technical strategy with business goals, setting and implementing technical principles, addressing scaling challenges, making informed decisions about new technologies, and fostering collaborative technical practices and decision making.

We emphasize the importance of using metrics for continuous improvement, introducing you to some useful metrics and teaching you how to apply them to assess organizational performance. You'll also learn how to develop capabilities to measure, test, and accelerate innovation through continuous experimentation, observability, comprehensive testing strategies, and balancing quantitative and qualitative feedback.

Throughout the book, we share real examples from our experiences, along with actionable tips and practices you can try. These examples will provide concrete illustrations of the concepts discussed, making them more relatable and accessible. By the end, you'll feel more confident and capable of leading through the chaos, equipped with practical strategies and insights that you can implement immediately.

You can't control the whole organization, but you can start building capabilities designed to enhance control and improve the situation. By focusing on what you can influence directly, you can create pockets of stability and efficiency that can serve as models for the rest of the organization. This approach enhances your immediate environment and demonstrates the effectiveness of your methods, potentially inspiring broader organizational change.

The book also includes appendices and resources, such as additional tools, exercises, extended reading, reference materials, and case studies to further your learning and application of the presented concepts.

The most successful engineering leaders we know didn't succeed despite the chaos; they succeeded because of it. Chaos strips away pretense, accelerates learning, and rewards those who can create order from disorder. While established companies debate decisions in committees for months, you can implement solutions in days. While they manage politics, you solve real problems. While they protect the status quo, you build the future.

By synthesizing the skills and strategies covered throughout the book and engaging in practical applications through simulated projects and real-world scenarios, you'll learn how to effectively lead in chaotic environments and drive success for your team and organization.

Conclusion: From Chaos to Clarity

You've seen the symptoms: a lack of ownership, constant firefighting, a blame culture, and teams shipping code without understanding what and why they are shipping. You may have recognized your own organization in these descriptions. Maybe you're feeling overwhelmed by the scale of dysfunction that you're facing. Here's the reality: you can't fix everything. The chaos won't magically disappear because you've identified it. But you can control your own approach to leadership—and that starts with getting clear on what your job actually is.

Most engineering leaders in chaotic environments fail because they try to do everything at once. They juggle fixing processes, mentoring engineers, managing emergencies, and setting strategy, all without a fundamental framework for their role. They end up exhausted and ineffective, contributing to the very chaos they're trying to solve.

The alternative is to focus on five core pillars that define engineering leadership:

People
> Building and developing your team, from hiring to performance management to creating a collaborative culture

Mission
> Connecting your engineers to the "why" behind their work, even when company direction is unclear

Plan
> Creating roadmaps and priorities that balance business needs with technical reality

Process
> Establishing workflows that enable rather than obstruct, keeping things lightweight but effective

Product
> Ensuring your team delivers outcomes that matter, not just output that looks good on a dashboard

These aren't abstract concepts; they're the practical foundations of your role. Get these right, and you create stability in your immediate sphere of influence. That stability then becomes a model for the broader organization.

In Chapter 2, we'll work through each pillar systematically. You'll assess where you are, define where you need to be, and create a concrete role description that reflects your actual responsibilities (not the generic job posting that got you hired). This isn't about creating more bureaucracy—it's about giving yourself a framework to operate from when everything else is in flux.

The chaos will still be there tomorrow. But at least you'll know what you're responsible for and how to approach it. Sometimes that's enough to get started.

CHAPTER 2

Understanding Your Role

In Chapter 1, we dove into the chaos, where unpredictability dominates and engineering leaders like you are tasked with navigating the storm. The stakes are high, but so are the opportunities if you can guide your team through the turbulence.

But before leading anyone, you must be clear on *your role*. In chaotic environments, things move fast, plans shift, and roadmaps are often missing. Your job is to create clarity, build stability, and keep your team anchored, even when everything around you is shifting.

It's not just about coding or managing tasks; it's about *setting direction* when none exists, creating cohesion out of chaos, and ensuring your engineers are connected to the bigger picture. And sometimes, even that picture is a little blurry.

In this chapter, we'll define what your role really is. You'll map out your core responsibilities and learn how to lead effectively when the rules keep changing. Consider this your onboarding guide to the role. Whether you're at a scrappy startup or a rapidly evolving enterprise, the foundation of your leadership remains the same: your people, your mission, your plan, your process, and your product.

What's Your Job, Exactly?

When you leaped into leadership, it might have felt simple: lead a team and ship great code. Easy, right? But here's the reality: leading a team is far more than meeting deadlines. You'll wear a thousand hats in chaotic environments, often on the same day. You'll be a people leader, a technical guide, a strategist, a product manager, a firefighter, and sometimes a therapist. The real challenge isn't just balancing these responsibilities; it's knowing when to wear each hat and how to keep everything moving forward.

Think about it like this: your role as an engineering leader is the glue that holds everything together. You're ensuring that your team isn't just working in isolation but building something that matters. And here's the secret: your team will mirror you. If you're confused, they'll be lost. If you're clear, they'll find clarity too.

The good news is that you don't need all the answers. You just need to build the framework, the core pillars that ground your leadership. We will break this down into five critical areas that define your job as an engineering leader, each essential to creating order from chaos: *people*, *mission*, *plan*, *process*, and *product* (as shown in Figure 2-1). We'll use each of these pillars to address the symptoms of chaos we described in Chapter 1.

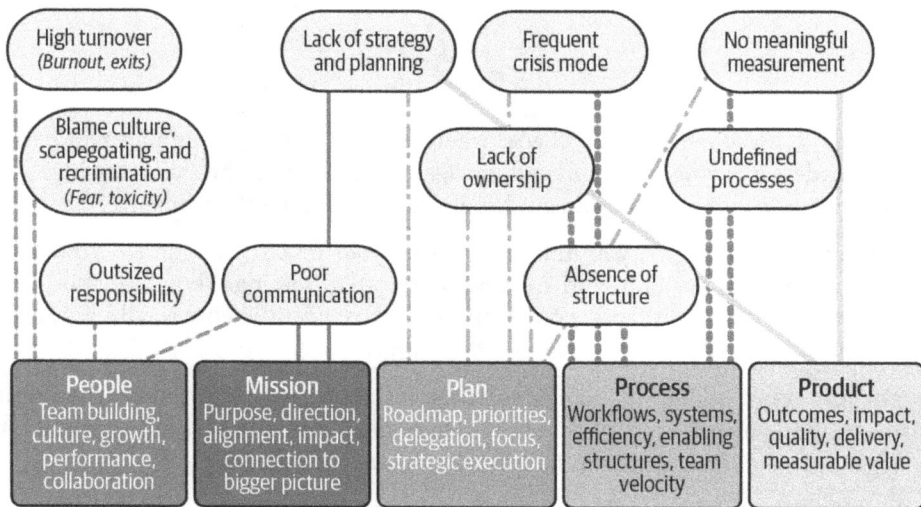

Figure 2-1. From chaos to clarity (a symptom–solution framework)

Table 2-1 summarizes the critical areas shown in Figure 2-1.

Table 2-1. Leadership intervention framework

Area	Intervention approach
People	Addresses high turnover, blame culture, outsized responsibility, and poor communication
Mission	Tackles lack of strategy, poor communication, and disconnection from purpose
Plan	Resolves lack of strategy, ownership issues, crisis mode, absence of structure, no measurement
Process	Fixes undefined processes, absence of structure, lack of ownership, frequent crises, and no measurement
Product	Establishes meaningful measurement, outcome focus, quality standard, delivered value, and strategic alignment

The Five Pillars of Engineering Leadership

At its core, your role can be distilled into five key pillars. Master these, and you'll have the foundation you need to thrive, no matter how chaotic your environment is:

People
> Everything starts with your team. Without them, you're not leading anyone. But it's not just about managing a headcount. You're responsible for building a culture, supporting growth, and ensuring that engineers work together as a unit, not just as individual contributors.

Mission
> Why is your team building what they're building? The mission is the "why" that drives your team forward. It's your job to connect the dots between what your engineers are doing and the bigger picture—how their work ties into the company's goals and what their impact is.

Plan
> Once the mission is clear, you must figure out how to get there. This is where you lay out the roadmap, prioritize work, and delegate effectively. Rather than dictating every detail, a plan should be about giving your team the direction and focus they need to move forward confidently.

Process
> A great plan is only helpful with the right processes to keep things moving. Process is what enables your team to work efficiently and stay on track. But here's the thing: ineffective processes will destroy progress. Your job is to set up processes that support your engineers, not bog them down.

Product
> Finally, consider what you are delivering. It's not enough to ship features; you must ensure that your team delivers outcomes that matter. Whether it's a feature, a fix, or an internal tool, everything your team does should have a measurable impact.

Boss, Where's the Roadmap?

So, how do we determine where we stand? Let's face it: most of us didn't step into leadership with a neatly defined job description that maps out every responsibility and tells us precisely what to do. More often than not, we're handed a vague mandate to "lead the team" and are left to figure out the rest.

But the truth is, even if you had a detailed job description, it probably wouldn't survive the first month on the job. Why? Because leadership, especially in chaotic environments, isn't static. Your responsibilities will shift and evolve as the organization grows—priorities change and new challenges will emerge.

That's why defining your role is so critical. You need a living document that outlines your responsibilities and how you plan to fulfill them—an adaptable document, just like you.

Don't Retreat to the Comfortable

Here's the temptation: when chaos arrives, you might want to retreat into what's comfortable or, worse, be frozen with inaction. For many of us, that means diving back into the technical work—refactoring or fixing bugs ourselves—because that's what we know. But the hard truth is that this is not your job anymore. If you're constantly doing the work of individual contributors, you're not leading; you're hiding.

As an engineering leader, your value is in how well you empower your team to succeed. And that means letting go of the need to be the technical expert in every situation. Your job is to guide, to enable, and to steer, rather than to do everything yourself. Leading engineers is not a promotion; it's a whole new job.

Engineering Leadership Is About Being a "Generalist with Range"

You've probably heard the phrase "jack of all trades, master of none," and maybe even felt a little queasy about it. But that's precisely what you need to be in an engineering leadership role. You don't need to be the world's best coder, nor do you need to be a product strategy expert. What you need is to have a little bit of knowledge about many things and the ability to move between them effortlessly.

In his 1973 novel *Time Enough for Love*, Robert A. Heinlein said (*https://oreil.ly/2gru5*), "Specialization is for insects." And it's true, while engineers are often specialists, leaders must be generalists. You need to be able to step into different domains—strategy, people management, and product alignment—without getting bogged down in technical details.

But let's be clear: this isn't an excuse for shallow thinking. Being a generalist means you have a broad understanding of how everything fits together and the insight to know when to go deep. It's about context switching with purpose, not just skimming the surface.

Action Beats Inaction Every Time

In a chaotic environment, the risk isn't that you'll do the wrong thing; it's that you'll do nothing at all. There's a tendency to freeze up when you don't have all the answers, waiting for clarity that may never come. But here's the truth: taking action is (almost always) better than standing still.

One of the most powerful traits of a great leader is a bias toward action. Even if you're unsure about the outcome, moving forward is better than waiting around. In chaotic environments, decisions are often made with incomplete information, and that's okay. It's better to make a decision, adapt, and iterate than to remain stuck.

And here's the good news we talked about in Chapter 1: chaos creates opportunity. Taking action, especially when no one else is willing to, sets you apart and builds momentum for your team. Don't be afraid to seek out opportunities to lead, even when the path isn't clear.

Role Description: A Do-It-Yourself Blueprint

So, how do you make sense of it all? How do you take control of the chaos and define your role in a way that brings clarity, focus, and direction to your team? It all starts with documenting your responsibilities. This is one of the first "hard parts" of your new role: communicating (in this case, writing well).

In the following sections, we'll break down the five foundational pillars: people, mission, plan, process, and product. As you work through each section, you'll have the opportunity to reflect on your current situation and define specific responsibilities that map to your unique challenges. By the end of this chapter, you'll have a clear, personalized role description that guides your leadership and helps you stay grounded, even when everything else is shifting around you.

Rather than being static, this blueprint is a living document that you'll refine and adapt as you grow. Whether you're building a team from scratch, inheriting a fractured group of engineers, or guiding a well-established team through turbulent times, this framework will give you the structure and clarity you need to lead effectively.

The first pillar is people. Before you can deliver a product, ship features, or implement processes, you need to understand the team that will build it all. Leadership starts with the people in front of you.

It's a common thought: "I've already got a job description and a team charter. Why do I need to define my role again?" Well, let's be real; most job descriptions are just glorified ads—aspirational and often out of touch with the reality of your day-to-day work. They're usually written before you even step into the role, and while they might look nice on paper, they rarely survive first contact with the actual job.

And charters? They often suffer the same fate—vague, filled with boilerplate language, or created in some long-forgotten workshop and never updated. Roles evolve, responsibilities shift, and those documents rarely reflect the new challenges you're taking on.

The real question is: *does your job description reflect what you actually do?* If not, having something more grounded can be a game changer. It gives you a clearer sense of direction and a tool to help you track your progress when it's time to showcase what you've been accomplishing.

People: It All Starts with Your Team

You've got the title, the responsibilities, and the office (maybe). But leadership? That starts with your people. No team means no leadership. It's not about getting them to do the work; it's about building something more significant than any individual effort. Your job is to ensure that your team isn't just a collection of talented engineers but a crew creating something extraordinary together.

But where do you start? Maybe you've inherited a team with a history you didn't write, or you're in the enviable position of building one from scratch. In either case, leadership starts with listening and seeing clearly. Don't get lost in the noise of missed deadlines, feature churn, or tech debt. The first task is always understanding your people—what makes them tick, what holds them back, and how you can help them grow into something stronger.

What's Your Starting Point?

Every team comes with a backstory, and before you start leading, you need to figure out which chapter you're in. Are you stepping into a well-oiled machine or taking charge of a team struggling with miscommunication and missed opportunities? Maybe they've been worn down by constant pivots, or perhaps they're a group of rockstars who only play solos, and don't collaborate. Wherever they are, your first move is to understand the dynamics already at work.

Take a good look at who is on the team. Who are the key players, and where are the gaps (both in terms of skill set and cohesion)? Are they communicating with each other or working in their own silos? Is there a culture of trust and collaboration, or does every meeting feel like a tug-of-war? These are the questions you need to ask before you start thinking about the future.

And if you're building from scratch, lucky you. You're not just hiring—you're shaping the foundation of your team. What kind of culture do you want to build? Do you want scrappy problem solvers, or are you looking for systematic thinkers who dot every "i" and cross every "t"? And how will you balance the team's seniority, experience, and fresh perspectives? Make no mistake: every hire and every decision you make at this point sets the trajectory for how your team will grow, scale, and deliver.

Then comes the key question: how much influence do you really have? Maybe you've been given complete control over team structure and hiring, or maybe you're working within constraints you can't change. Understanding those limits is just as important as recognizing your team's potential, because knowing what you can't change will help you focus your efforts where they matter most.

What's Your Mission?

Here's the reality: your goal isn't to micromanage people or simply fill a headcount quota. You're building a team that's greater than the sum of its parts. A truly aligned team doesn't just collaborate; they flow. They're meeting objectives without needing endless meetings or follow-ups because they understand the mission and their role in achieving it. That's where you're headed.

The question is: does your team understand why they're here? Are they just coding to ship features, or do they understand how their work ties into the bigger picture? Engineers need to see the impact of their efforts—how their code affects the product, how that product serves users, and how it supports business outcomes. Without that understanding, even the most skilled developers can lose steam. Your job is to make sure that never happens. When your team understands the "why," they stop merely working and start innovating.

Software development is social. No matter how talented an individual may be, the interplay of communication, collaboration, and feedback drives exceptional results. If your team is missing that, it requires more than process adjustments; it calls for a culture where collaboration is second nature. When you achieve that, your team delivers meaningful results.

What Tools Do You Need to Get There?

Leading isn't about hand-holding, nor is it about just keeping track of deadlines. It's about creating an environment where your team can thrive. And that starts with knowing where they stand, especially regarding morale. If your team is disengaged or burned out, no amount of technical skill will keep them productive. You can't fix morale with snacks, shout-outs, or vague promises of "we'll do better." It takes real leadership: start with small, meaningful wins and honest conversations. Acknowledge pain points, and then establish a clear and achievable path forward.

Next, consider performance management. Does your organization have systems in place, or is it your responsibility to create structure? If your company lacks support, you're still accountable. You need ongoing feedback loops to help your engineers grow, rather than relying solely on annual reviews to course correct. Performance management is about guidance, not evaluation.

Finally, let's talk about hiring. If roles are open, how much influence do you have over who joins your team? Are you actively shaping the hiring process to ensure candidates fit both technically and culturally, or are you simply filling seats? Once hired, what's the onboarding plan? The first 30 days set the tone. If your new hires don't feel supported and integrated from day one, you're looking at months of playing catch-up. It's not enough to just bring people in; you need to set them up for success right from the start.

Exercise: Defining Your Role as a People Leader

Now that you've thought deeply about your team, it's time to translate those insights into action. This is the first step in defining your leadership role, starting with the foundation: your people. We'll break it down step by step, helping you assess where your team is now, where you want them to go, and what specific responsibilities you'll take on to make it happen. Don't rush through this. How you lead your people shapes everything else that follows.

As you complete the exercise, you'll sketch out the foundation of your leadership role: the people. This is the first, and arguably the most important, piece of the puzzle.

Step 1: Assess Your Team's Starting Point

Take a moment to reflect on where your team stands today. Write two or three sentences that summarize the current state of your team. Are you inheriting a team or building one from scratch? How engaged are they? What's working, and what's not? Example answer:

> I've inherited a technically solid team, but the energy is low, and collaboration is almost nonexistent. Communication feels like an afterthought, and deadlines have been slipping for months. I have room to hire, but the current structure is unsustainable.

Step 2: Visualize Your Team's End State

Now that you've diagnosed the present, it's time to dream about the future. What kind of team are you trying to build? Be specific, don't just say you want a high-performing team. What does that mean? Think about the balance between collaboration, technical expertise, and innovation. Write two or three sentences that outline your vision. Example answer:

I want a team that not only ships quality software at speed but also deeply understands the business impact of their work. I'll focus on building a collaborative culture where communication happens naturally, and senior and junior engineers feel empowered to contribute meaningfully.

Step 3: Sharpen Your Role into Specific Responsibilities

It's time to get practical. What are the specific things you will do to lead this team from where they are now to where you want them to be? Write 5–7 responsibilities that reflect your key actions as a people leader. Think about how you'll manage performance, foster growth, and support collaboration. Make these responsibilities actionable and clear.

For this task, consider April, who has just taken over the StudioOps team at PixelCurl. It's a team with incredible talent but fractured collaboration. Senior engineers often work in silos, leaving junior engineers under-mentored and struggling to succeed. The team is demotivated, and deadlines continue to slip. April's challenge is to rebuild team cohesion, improve mentorship, and strategically hire to balance skills gaps.

Example answer in April's context:

- *Oversee StudioOps performance, ensuring collaboration is woven into daily operations and silos are broken down. April knows her top priority is fostering openness between senior and junior engineers to encourage more cross-pollination of ideas.*

- *Mentor engineering managers, guiding them to step into more effective leadership roles. April needs others to take on more people management responsibilities, enabling her to focus on long-term strategy.*

- *Identify skill gaps and recruit strategically, hiring new engineers to balance the team's capabilities. April is hiring three new team members to fill critical gaps in frontend development and DevOps, ensuring the team is well rounded for future growth and success.*

- *Develop a structured onboarding process that helps new hires feel integrated from the start. In the past, new hires have struggled to find their place, and April wants to change that by creating an onboarding experience that sets them up for success.*

- *Cultivate a culture of continuous learning, where technical excellence and business acumen go hand in hand. April's goal is to help her team connect the dots between their code and the real-world problems they're solving for PixelCurl's users.*

We'll see more of April throughout the book, and we'll share more about building your team in Chapter 4.

Mission: You've Got a Crew; Now You Need a Place to Go

You've got your team assembled. They're working, shipping code, and pushing through "sprints." But the question is, do they know why? What are they working toward? And more importantly, do they care?

Your job isn't just to manage projects or move features across a Kanban board; it's to connect your team to a mission bigger than themselves. Engineers thrive when they understand what they're building and why it matters. They need to see the bigger picture: how their work shapes the product, impacts users, and drives the company forward. Without that connection, even the most talented team will drift. We, as leaders, solve problems. So, the goal is to get your team closer to the problems!

You might be wondering, what if the mission isn't clear from the top? What if the company's goals are shifting or vague? That's where you come in. Your role is to create clarity where there isn't any, giving your team a sense of purpose even when the road ahead is dark and foggy. The mission might not always be perfectly defined by leadership, but that doesn't mean your team can't have its own mission. Whether it's maintaining technical excellence, driving innovation, or solving tough user problems, you must give your engineers something concrete to rally around.

Where to Start

Before you can get your team aligned with the mission, you need to assess where things stand. Does the company have a well-communicated mission that everyone understands, or are your engineers stuck in the weeds, coding features without seeing how their work connects to broader goals? Even worse, are they choosing technologies simply because they are interesting, rather than because they serve the mission?

If your team isn't already connected to the company's mission, that's your starting point. But don't expect a corporate vision statement alone to inspire them. Your job is to turn that high-level mission into something tangible. Engineers need to understand how their work contributes—whether by improving the user experience, driving product adoption, or supporting key business goals. It's not enough for them to know the mission exists; they need to see how their daily efforts make a measurable impact.

If the company's mission is unclear or changes frequently, you need to step in and create clarity for your team by translating the company's priorities into something your team can act on. It may be about establishing stability in the product or reducing friction for users. Whatever the goal, ensure your team has a clear, actionable direction that aligns with the company's broader objectives, even if those objectives feel a bit uncertain.

Getting Your Team on Board

Once you've clarified the mission, the next challenge is getting your team to care about it. It's easy to assume that engineers are aligned just because they're doing the work, but real alignment goes beyond completing tasks; it requires personal connection.

Your team needs to feel that they're not only delivering features but solving meaningful problems. And that sense of purpose doesn't come from the top down alone; it comes from you, bridging the gap between technical work and real-world impact. Help them see how that new API they're building will improve the customer experience, or how an infrastructure upgrade is laying the groundwork for future scalability.

The key is to show them that their work has tangible value. When engineers understand that what they're building is a solution to a real problem, they become more motivated, creative, and invested. They stop thinking of themselves as just executors of tasks and start seeing themselves as contributors to something larger. And that's when the magic happens—because a motivated, mission-driven team often exceeds expectations.

Turn the Mission into Action

So, you've got the mission. But how do you make it real? A vision is useless without action, and your team needs more than high-level goals; they need concrete, measurable objectives that make the mission tangible in their day-to-day work.

If you're responsible for setting goals, ensure they align with the company's mission and are specific enough to drive action. For example, rather than just aiming to "improve performance," tie it to something meaningful: "Reduce customer onboarding time by 20%." Instead of focusing on "reducing technical debt," consider linking it to "increasing uptime to support enterprise clients." This is how you turn the mission into something your engineers can sink their teeth into. When they see how their work makes an impact, they'll be more engaged, focused, and driven to push the boundaries of what's possible.

And if your company's mission is unclear, don't wait around for direction; go get it. Connect with the product, marketing, and leadership teams. Understand the broader business goals and bring that clarity back to your team. Beyond just managing code delivery, your job is to create a bridge between business objectives and technical execution. In doing so, you will be ensuring that your team's work is aligned with the company's strategy, even when the roadmap is constantly evolving.

Exercise: Mission Alignment with Measurable Impact

This exercise will help you to define concrete, measurable approaches to connecting your team with the company's mission. Instead of vague alignment goals, you'll create specific indicators that show whether your team is genuinely connected to meaningful work.

Step 1: Diagnose Current Disconnection

Pick one symptom from Chapter 1 that your team is experiencing. Then identify how disconnection with the company's mission might be contributing to it. Example answer:

> *Our team has high levels of context switching and unclear priorities. I think this happens because we take on any urgent request without understanding what projects actually matter most to the company's success.*

Step 2: Define Your Measurable Outcome

What would change if your team truly understood how their work connects to company success? Write one specific, measurable change you'd expect to see in 90 days. Example answer:

> *Engineers are able to explain in their own words how their current project connects to company goals, and we spend less than 20% of our capacity on unplanned work that doesn't align with strategic objectives.*

Step 3: Create Your Action Plan

List two to three concrete actions that you'll take to achieve this outcome. Make them specific enough that someone else could execute them. Example answer:

> - *Hold monthly impact sessions where we trace infrastructure improvements to customer outcomes (faster deploys, lead to quicker feature delivery, which lead to faster user feedback loops).*
> - *Conduct quarterly roadmap reviews with business leadership to ensure platform priorities align with company strategy.*
> - *Track and share metrics showing how platform improvements enable feature team velocity.*

Step 4: Pick Your Leading Indicator

How will you know within 30 days whether this approach is working? Choose something you can measure quickly. Example answer:

> *In team meetings, engineers start connecting their work to business outcomes without prompting, and feature team requests include business justification rather than just urgency claims.*

The goal is to connect daily work to meaningful outcomes in ways you can actually measure, rather than just hoping for cultural change that might never materialize.

We'll talk more about setting direction and technical strategy in Chapters 5 and 9.

Plan: You Know Where to Go; Now Chart the Course

It's one thing to know where you're headed, but it's another thing entirely to map out how to get there. While a mission gives your team a destination, what about the plan? Without a roadmap, even the most talented engineers can get lost. A plan should do more than just list tasks; it must create focus, alignment, and momentum.

But here's the twist: you're not expected to create the plan alone. The best plans are built collaboratively, bringing in insights from all directions. That means liaising with your cross-functional partners in product, design, and marketing teams, and empowering your technical leads to take ownership of parts of the roadmap. A leader who plans alone is destined for bottlenecks, but a leader who plans with others sets the stage for execution at scale.

Build with Your Partners

If you think you can map out the perfect plan on your own, you're probably setting yourself up for failure. As a leader, your job isn't to have all the answers; it's to ask the right questions and bring in the right people. Rather than being created in isolation, great plans result from collaboration across functions, with product managers who understand customer needs, marketing teams that see market shifts, and sales teams who comprehend customer pain points.

Aligning engineering's efforts with the company's broader goals requires an in-depth understanding of business priorities, product vision, and customer impact. So start by building strong relationships with your partners in those departments. Sit down with the product team to understand what's coming down the pipeline, and connect with sales to learn what's breaking the deal. The more context you have, the better you can translate business needs into engineering goals.

And don't forget about your technical leads. The engineers closest to the code often have the best sense of where things could go wrong or where potential opportunities lie. Bring them into the planning process and let them own pieces of the roadmap, so that they become leaders in driving the plan forward.

Know What to Prioritize

A plan is only useful if it helps your team focus on what matters most. When everything is important, nothing is. This is where the magic of prioritization comes in. The challenge isn't figuring out what your team *can* do; it's deciding what they *should* do right now to deliver the most value. And remember: it's not prioritizing if it doesn't hurt. If you're not making tough calls, you're not really prioritizing.

So ask yourself: what's going to move the needle the fastest? Should you focus on shipping features that drive growth, or is your tech debt piling up to the point where it's slowing everything down? Is your infrastructure holding back scalability, or is there a feature that could unlock new customer adoption? The answer won't always be obvious, and it won't always stay the same. But as a leader, you need to be decisive.

In prioritization, sometimes, saying no is more important than saying yes. Roadmaps and backlogs are filled with good ideas, but you should only pursue the most relevant and important ones. Saying "not today" is usually challenging, but remember that a "no" today is not "never"; it's just not today. Your team needs clarity, which means making those tough decisions. When you clearly communicate the reasons behind those decisions, your engineers will be confident to move forward without second-guessing. They'll understand not just *what* they're doing but *why* they're doing it.

Turn Strategy into Action

A plan is worthless if it doesn't get executed. To move from strategy to action, break down high-level goals into actionable steps that your team can rally around. It's not enough to say, "We need to improve performance." What does that look like in practice? Does it mean upgrading the architecture, refactoring old code, or optimizing the database? Every big goal needs to be distilled into specific, measurable objectives.

It's important to delegate responsibilities, rather than attempting to execute alone. Trust your technical leads to take ownership of key parts of the plan. Instead of micromanaging, your job is to steer the ship. Trust your leads to handle the details while you keep an eye on the horizon, ensuring the team stays aligned with the bigger picture.

It's also important to build checkpoints into your plan. Don't wait until the project's done to see if things are on track. Set milestones along the way—clear markers that enable you to assess progress and course correct if needed. These checkpoints guide your team, letting them adjust without losing sight of the end goal.

Adjust, but Stay Focused

No plan survives first contact with reality. Things will change; priorities will shift, new information will come to light, and unexpected obstacles will arise. Great leaders know how to stay flexible without losing focus. When the roadmap changes, it's your job to keep the team grounded. The core objectives don't change, even when the tactics do.

This doesn't mean constantly reworking the plan, but rather staying nimble and reevaluating when necessary. Maybe a new opportunity has opened up that's too good to pass up, or maybe product feedback has shifted a feature's priority. When things change, don't panic—adapt. But always ensure your team understands how these changes fit into the broader mission. They must feel like they're still moving forward, even when the path shifts.

Exercise: Defining Your Role in Strategic Planning

Now that you've explored how to create a plan that drives focus and alignment, it's time to turn that into actionable responsibilities for your leadership role. This exercise will help you define how to collaborate, prioritize, and ensure that your team stays on track.

Step 1: Assess Your Current Planning Approach

Take stock of how planning is done today. Is there a clear roadmap guiding your team's efforts, or is everyone just reacting to the next urgent request? Write two or three sentences about your team's current state of planning. Example answer:

Our planning process is reactive, driven by the product team's last-minute requests. We lack a long-term roadmap, and my engineers constantly shift gears without a clear sense of priorities.

Step 2: Define Your Approach to Planning

Next, consider how you can develop a more structured and collaborative planning process. Write two or three sentences about how you'll bring key partners into the process and set clear priorities. Example answer:

I'll work closely with product and marketing to align engineering efforts with business priorities. I'll also involve my technical leads in shaping the roadmap, giving them ownership of key initiatives to drive accountability and buy-in.

Step 3: Add to Your Role Description

Now, outline two to three responsibilities that define your role in leading the planning process. Think about how you'll manage collaboration, set priorities, and translate strategy into actionable steps for your team.

Example answer in April's context:

At PixelCurl, April's team has been stuck in a reactive cycle; constantly shifting focus based on the latest product demands. There's no long-term roadmap, and the engineers are feeling the strain of constant change. April needs to develop a plan that aligns with the company's goals and provides her team with clarity. April could outline the following responsibilities:

- *Collaborate with product, marketing, and operations teams to ensure engineering efforts are aligned with business objectives. April must build stronger relationships with key partners to create a cohesive plan that drives company success.*

- *Set clear priorities that balance short-term feature delivery with long-term technical improvements. April knows that focusing solely on feature velocity won't suffice. Her team must balance immediate product needs with addressing critical technical debt and scalability challenges.*

- *Delegate execution to technical leads, empowering them to own key pieces of the roadmap. By giving her leads responsibility for driving initiatives, April creates accountability and space for her to focus on guiding the overall strategy.*

With this exercise, you'll have defined how to approach planning—from collaboration to prioritization to turning strategy into actionable steps. A clear, well-thought-out plan is the backbone of execution, ensuring your team knows where they're going and how to get there.

We'll talk more about planning and technical strategy in detail in Chapters 6 and 8.

Process: The System That Keeps Things Moving

You can have the best team, the clearest mission, and the smartest plan, but if your processes are a mess, then none of it matters. Process is the engine that powers execution. Done right, it keeps your team efficient, focused, and productive. Done wrong, and it becomes the bottleneck that grinds everything to a halt.

So, what is the key to great process? Less is more. You don't need to create an elaborate system of approvals and checkpoints, as this leads to bureaucracy. Instead, what you need is a framework that enables work to flow smoothly without getting in the way. The best ones are there to support, not to control.

Keep It Lean, but Effective

When it comes to processes, complexity is the enemy. If your team is bogged down by long approval cycles, endless meetings, or cumbersome code reviews, they'll lose momentum and motivation. Great processes are lightweight and designed to remove friction rather than add it.

But simplicity doesn't mean a free-for-all. You still need structure, code reviews, deployment processes, and feedback loops, but these should be streamlined to keep things moving. If a process is dragging, take a step back and ask, "Is this helping us get better, or is it slowing us down?" If it's the latter, cut it.

Make Process a Team Effort

You're not the only one responsible for defining the process. The best processes are created with or by the team, not for the team. Your engineers are the ones who have to live with these processes every day, so it only makes sense to involve them in shaping how things are done.

This isn't just about efficiency but about ownership. When your team has a hand in creating the process, they're far more likely to follow through and keep things running smoothly. Create space for feedback and be open to adjustments when things aren't working. Process is a living system, not a static rulebook.

Adjust As You Go

No process is perfect forever. What worked when your team consisted of 5 people might fall apart now that you're a team of 20. Processes need to evolve as your team grows and the complexity of your work increases. That means regularly evaluating how things are going—considering what's working, what's slowing the team down, and where improvements can be made.

Don't be afraid to change what's not working. Just because a process was useful last quarter doesn't mean it's relevant now. Keep things flexible, and don't hesitate to scrap something if it's no longer serving the team.

Exercise: The Process Pulse Survey

Most leaders guess at process problems instead of measuring them. Your team knows exactly where the friction is; you just need to ask the right questions. In the end, they are the ones who consume your "process." Listen to your users.

Deploy this two-minute anonymous survey to track team productivity patterns:

1. This week, how often did you get to work uninterrupted for 2+ hours?
    a. Most days
    b. A few times
    c. Once or twice
    d. Not at all
2. How often do you find unofficial ways to get things done?
    a. Never—official processes work fine.
    b. Occasionally—for edge cases.
    c. Regularly—it's faster.
    d. Always—it's the only way.

---

3. What do you spend the most time waiting for?

   a. Reviews or approvals

   b. Other people's work

   c. Access or permissions

   d. Decisions from above

   e. Information or clarification

   f. Nothing significant

4. For the work you do daily, how quickly can you usually get the go-ahead?

   a. Immediately

   b. Same day

   c. One or two days

   d. A week or more

5. What would you tell a new teammate if they asked, "How do things really work around here?"

6. Complete this sentence: "If I could change one thing about how we work, I would…"

Reading the Results

Start by *listening*, not reacting. The "process pulse" survey reveals how work actually flows, not how you think it does. Look for patterns. Are people waiting on approvals, inventing workarounds, or struggling to find focus time? These are signals of friction, not complaints. The open-ended responses often reveal the real story, including the unspoken rules, backdoor processes, and quiet trade-offs that shape your culture. Together, they show how work truly gets done.

Iterating Toward Fixes

Be thoughtful about how you try to address the issues. Most leaders try to fix everything at once, and nothing changes. Instead, focus on one thing. Find the single pattern that slows down the most people the most often, and design a two-week experiment to improve it.

Keep it small and fast. The goal is progress, not perfection. Here are some possible experiments:

- Approve routine changes the same day.
- Protect two-hour focus blocks each morning.
- Create a fast lane for small, low-risk requests.
- Simplify deployment steps to remove waiting.

These small shifts send a signal: improvement is possible, feedback matters, and bureaucracy is not the default.

Measuring Impact

Rerun the pulse after two weeks. Look for changes in both metrics and mood: more flow time, fewer workarounds, faster decisions, and a working environment that feels lighter.

Share what you learn. Celebrate progress, acknowledge what did not work, and keep the loop going. Trust grows when teams see their feedback turn into action, and over time, that trust becomes the engine of continuous improvement.

We'll talk more about processes and collaboration in Chapters 6 and 9.

Product: What Are We Actually Delivering?

At the end of the day, it's all about results. Your team might be pushing code, but are they delivering outcomes that drive the business forward? In engineering, it's easy to get caught up in shipping features, closing tickets, and hitting deadlines. But all that means nothing if what you're delivering doesn't solve real problems or create user value. Outcomes matter more than output.

This is where leadership steps in. Your job is to make sure the work your team is doing translates into real-world impact. Whether that's a customer-facing feature, a critical infrastructure upgrade, or an internal tool that boosts productivity, your team's work needs to move the needle. The best leaders don't just focus on getting things done; they focus on getting the *right* things done. And that's the difference between a team that just releases features and a team that delivers value.

Know What Success Looks Like

Your team will chase the wrong goals if you don't define success. And hitting a deadline or closing a sprint doesn't always mean success. The real question is: what problem are you solving? Are you building something that genuinely makes a difference to your users or the business, or are you just shipping for the sake of shipping?

It's your job to guide your engineers toward the impact of their work, not just the output. Every feature and every line of code should be tied to a clear outcome. Are you improving user experience? Are you making the product faster, more reliable, or easier to use? Are you enabling the company to scale, grow revenue, or reduce costs? When your team can see the impact of what they're building, they'll care more, innovate faster, and take pride in their work.

In reality, success is fluid. It might be releasing a new feature this quarter, and it could be driving down infrastructure costs next quarter. Success is about delivering value, and values change as the business evolves. Your role is to keep that focus clear, even as priorities shift.

Balance Quality with Speed

Speed is critical. Shipping fast gets you ahead of the competition, unlocks new opportunities, and helps the company grow. But speed without quality? That's a disaster waiting to happen. Too often, teams push to ship faster, only to cut corners, leading to tech debt, bugs, and dissatisfied users. Shipping fast isn't enough; you need to ship well.

Your job is to balance those two forces. Yes, you want your team to move quickly, but not at the expense of building something that doesn't work. The trick is figuring out how to maintain velocity without sacrificing quality. This might mean tightening up your testing processes or setting clear standards for what "done" looks like so that your team doesn't ship half-baked features.

But there's a nuance here: quality doesn't always mean perfection. It's about knowing when something is good enough to ship and when it still needs work. Great leaders help their teams find that balance, pushing for progress while ensuring what's delivered actually adds value. Your team needs to trust that when they ship, they're delivering something meaningful, not just ticking a box.

Outcomes Aren't Always Features

Here's a common trap: thinking that only the visible work matters. Sure, releasing a shiny new feature feels great, but that's not the only work that drives outcomes. Sometimes, the most important work your team does isn't something users will ever see. Infrastructure, technical debt, and internal tools are the silent engines that power the product.

It's easy for teams to lose motivation when they're working on backend improvements or technical debt that doesn't get flashy recognition. However, as a leader, your job is to help them see the bigger picture. Explain how this kind of work enables future growth, reduces bugs, and makes everything run smoother. The invisible work is often the most critical, and it's up to you to ensure your team understands its value.

It's about ensuring that the work done today sets the company up for success tomorrow. Celebrate these hidden wins and help your team see that they're building the foundation for the future.

Exercise: Redefining Your Role for Impact

In this final exercise, you will define your role as a leader in ensuring that your team delivers outcomes that matter. You will reflect on how to help your team focus on impact, maintain quality while moving fast, and make sure even the invisible work creates real value.

Step 1: Assess Your Team's Focus

Start by reflecting on your team's current reality. Are they creating meaningful impact, or just completing tasks? Write two or three sentences describing where your team stands today in terms of delivering value. Example answer:

My team ships code regularly, but they are more focused on hitting deadlines than solving real user problems. We move fast, but quality issues are starting to erode trust.

Step 2: Define Your Leadership Approach

Now describe how you will shift your team's focus from output to outcomes. Write two or three sentences about how you will help them balance speed with quality and make their work count. Example answer:

I will help my team connect their work to customer outcomes, not just tickets closed. We will build habits that protect quality while moving fast, and invest time in the invisible work that keeps our systems strong.

Step 3: Add to Your Role Description

Describe how you will define your role in leading for outcomes. These should describe what you will own as a leader, ensuring quality, managing pace, and reinforcing the value of all work, visible or not. Example answer in April's context:

At PixelCurl, April's team has been racing to ship features, but the rush to move fast has resulted in poor-quality code and rising customer complaints. April needs to recalibrate the team's focus, balancing speed with quality and ensuring that even the invisible work (like infrastructure improvements) is given the attention it deserves. April's role description could be as follows:

- *Define success metrics prioritizing impact over output, ensuring that her team's work drives real value for users and the business. April will guide her engineers to focus on delivering outcomes that matter, rather than just meeting deadlines.*

- *Balance speed with quality, making sure that while the team moves fast, they don't cut corners that lead to more problems down the road. April will push for velocity, but not at the expense of user experience or product stability.*

- *Emphasize the importance of invisible work, recognizing that technical debt, infrastructure improvements, and internal tools are essential components of success. April knows that a healthy product needs more than just new features; it requires strong foundations.*

By completing this exercise, you will have a clear statement of your leadership focus, guiding your team to deliver meaningful results, maintain high standards, and ensure every line of work adds lasting value.

We'll talk more about managing products and shipping in Chapter 6.

Conclusion: Pulling It All Together

This chapter gave you a comprehensive roadmap for defining your leadership role. From people to mission, planning to process, and finally, outcomes, you've dug into the heart of what it means to be an effective engineering leader. You've asked the tough questions, reflected on where you are and where you want to go, and translated those insights into an actionable role description.

But remember: leadership isn't static. As your team grows and your company evolves, so will your role. Continue revisiting these questions, refine your approach, and push yourself to lead and inspire.

In Chapter 3, we'll dive deep into the diagnostic skills you'll need to start to navigate the chaos.

Navigating Chaos

You probably didn't step into this role because everything was going smoothly. Most engineering leaders are hired or promoted because something is broken; a team is behind; trust is shaky; delivery is inconsistent; morale is slipping; the roadmap is on fire—and now it's yours to fix.

There's no checklist for this moment, no onboarding guide for chaos. There is just pressure, ambiguity, and a team looking to you for direction.

Here's the part that most people won't admit (or aren't aware of): you don't fix this with control; you fix it with *range*. Figure 3-1 gives you an overview of what this means.

Range gives you the ability to read the room and adjust. It enables you to lead with clarity when it seems like there is none around you, to be decisive without being rigid, and to calm things down without slowing them down.

That's what this chapter is about. Leadership in chaos isn't about having the perfect answers; it's about knowing how to show up effectively and when to shift your stance so the team can move forward, even when everything feels stuck.

We call these shifts *switches*: deliberate moves in how you lead, tuned to the moment. Rather than being personality changes or performance tricks, they're adjustments in tone, pace, and focus so that your team gets what they actually need right now.

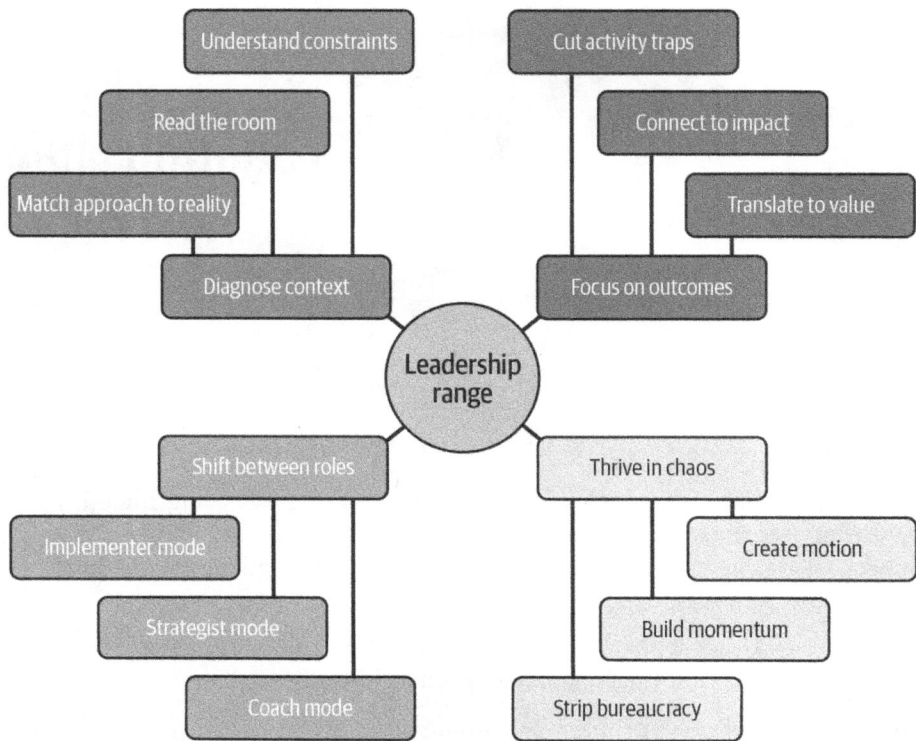

Figure 3-1. Breakdown of leadership range

This chapter walks through the four key switches that make up your leadership range, as identified in Figure 3-1:

Diagnose context
    Matching your leadership to the moment

Thrive in chaos
    Creating momentum when everything stalls

Focus on outcomes
    Steering attention toward impact, not just activity

Shift between roles
    Knowing when to be a pilot, medic, or engineer

Each switch is both a lens for reading the situation and a tool for acting with purpose. Together, they give you the range to lead through chaos.

Diagnose Context

Match your leadership to the moment. One of the fastest ways to lose a team is leading them as if they're somewhere else.

A startup crew under existential pressure doesn't need the same guidance as a scaled team optimizing for predictability. A burned-out group struggling through low morale needs a different touch than a hungry team racing toward a deadline.

Yet many leaders, especially in chaos, fall back on what worked before. The context has shifted, but their approach hasn't. That misalignment creates friction, wastes energy, and often makes problems worse instead of better.

Diagnosing context means understanding the real conditions before you act. When you match your leadership to the moment, your instincts become assets rather than liabilities.

Why Diagnosis Matters More Than Style

Context isn't just about structural differences between startups and enterprises, or small teams and large organizations. Effective diagnosis requires reading three distinct dimensions that shape how teams receive and respond to leadership:

Human dimension
> This refers to the emotional and psychological state of your people. Are they energized and ready for challenges, or depleted and needing stability? Are trust levels high enough to support risk-taking, or has skepticism created defensive behaviors?

Operational dimension
> This dimension is about the current state of work systems and processes. Are priorities clear and is work visible, or is the team drowning in hidden tasks and competing demands? Is the pace sustainable, or are people constantly putting out fires?

Cultural dimension
> The third dimension involves the unspoken rules about what behaviors are rewarded or punished. Does the environment encourage innovation or punish failure? What feels safe versus risky in this particular context?

Each dimension shapes how your team receives and responds to leadership. If you miss any of them, you'll find yourself pushing solutions that don't fit the problem.

Four Lenses for Reading Context

The best way to develop this awareness is by examining the same situation from different lenses. Like the dimensions used to diagnose your context, each lens reveals other aspects of what's actually happening, helping you choose responses that fit rather than defaulting to familiar patterns that might be entirely wrong for the moment.

1. The system lens: What kind of environment are we in?

Every team operates within larger constraints that set the pace, risk tolerance, and types of problems that dominate. If you don't account for these systemic pressures, you'll optimize for the wrong things and wonder why nothing works.

A startup team thrives on speed and improvisation, but the same behavior in a compliance-heavy environment creates risks that could kill the company. An enterprise team values predictability and careful coordination, but the same caution in a startup can be equally costly.

In startups, hesitation means missing market windows, losing first-mover advantage, and burning runway while competitors move ahead. Customer needs evolve rapidly, and the luxury of extensive planning often doesn't exist. Teams that spend months perfecting features may discover the market has shifted, funding has dried up, or a competitor has captured the opportunity. The cost of moving slowly can literally be business survival. Startups typically have limited time and resources to find product–market fit before running out of capital.

Enterprise caution that works in established markets becomes a liability when speed determines whether you capture emerging opportunities or lose them forever. In Table 3-1 we summarize how different organizational stages require different strategies for success.

Table 3-1. Different leadership approaches required by different organizational stages, with early-stage chaos requiring different skills than enterprise complexity

Environment	Defining traits	Leadership focus
Early stage	Existential pressure, resource constraints, rapid iteration	Protecting energy, simplifying decisions, maintaining focus
Scale-up	Growth outpacing systems, unclear ownership, increasing complexity	Building sustainable practices, clarifying boundaries, managing overwhelm
Enterprise	Established processes, political dynamics, interdependencies	Navigating stakeholders, influencing indirectly, finding leverage points

2. The human lens: What shape are my people in?

The same system can feel completely different depending on your team's emotional and mental state. Morale, trust, and energy levels determine whether people can absorb new challenges or need stability first. Table 3-2 highlights some of the warning signs and suggested responses for concerns in these categories.

Table 3-2. Team energy, trust, and history determining readiness for change—the importance of diagnosing emotional capacity before pushing forward

Factor	Warning signs	Leadership response
Energy levels	Signs of fatigue versus enthusiasm in meetings and work quality	Protecting recovery time or channel momentum appropriately
Trust dynamics	How freely people raise problems and disagree constructively	Rebuilding through small, visible commitments
Historical baggage	References to past failures, leadership changes, and broken promises	Acknowledging history, creating safety before pushing change

Here's the thing about burnout: it often masquerades as resistance. A team that seems disengaged might actually be exhausted from months of navigating chaos or organizational churn. Pushing harder in these moments deepens the problem rather than solving it. Therefore, it's important for leaders to distinguish between genuine resistance and capacity depletion before choosing their response.

Resistance shows up as active pushbacks and vocal concerns. Burnout masquerades as withdrawal, declining quality, and hollow compliance. People stop volunteering ideas and just go through the motions.

The interventions are opposite: resistance needs vision clarification and buy-in rebuilding; burnout needs capacity protection—slowing down, reducing scope, or creating recovery time before pushing forward.

You can have perfect strategy and culture, but if your people are running on empty, everything will feel like resistance. Sometimes leaders need to step back to move forward more effectively.

3. The work lens: What's actually on the plate?

Teams can appear dysfunctional when the real issue is invisible or misaligned work. Even high-performing groups struggle under hidden support requests, unclear priorities, or a difficult ratio of unplanned to planned work. Table 3-3 outlines some of these productivity illusions and possible solutions.

Table 3-3. Productivity illusions created by hidden work and scattered focus, and leadership actions to make work visible and protect sustainable pace

Work pattern	Warning signs	Leadership action
Hidden work	Complaints about being "busy" but with low visible output	Making all work visible, tracking interruptions
Low-impact focus	Lots of activity, but unclear business connection	Reconnecting work to outcomes, cutting low-value tasks
Unsustainable balance	Constant crisis response or endless optimization	Protecting capacity, enforcing prioritization boundaries

4. The cultural lens: What are the unspoken rules?

Culture shapes what people feel safe doing, what gets rewarded, and what provokes resistance. You can have the right strategy but the wrong cultural approach, and see your efforts get rejected like a bad organ transplant. Table 3-4 suggests leadership approaches based on culture shapes.

Table 3-4. Amplifying cultural strengths while managing liabilities

Cultural dimension	Cultural strengths to amplify	Leadership action
Speed versus stability	Move fast and break things / measure twice, cut once	Calibrating pace to context, communicating rationale for timing decisions
Decision-making style	Consensus-driven / top-down clarity	Setting decision timelines, creating safe dissent channels
Risk tolerance	Innovation rewarded / failure punished	Aligning incentives with desired risk behaviors, celebrating intelligent failures

The Reality of Power and Identity

Here's what most leadership advice doesn't tell you: context diagnosis isn't neutral. Your background, identity, and position within the organization influence what you can see and the actions available to you. You've probably felt it, but you have to lean into it.

Some leaders can challenge executives directly and walk away with increased trust. Others face higher scrutiny for the same behavior due to their background, accent, or visible identity markers. Leaders from underrepresented backgrounds frequently navigate additional political complexity that affects which leadership approaches are viable, and those holding multiple marginalized identities may face compounded challenges that shift based on context.

While this may not be fair, understanding your constraints will help you to choose strategies that work within your actual sphere of influence rather than an idealized version that exists only in leadership books that don't consider these factors.

We're not writing this from a neutral position either. As coauthors—both bald, tattooed men, one Colombian, one Australian—we've learned that while we may share the baldness, our boldness shows up in different ways. Accents get noticed and assumptions are made. Through this, we've learned that the leadership approaches that work for one of us might not work for the other.

It is important to keep this in mind when anyone gives you leadership advice—including us. What works in our contexts might completely miss the mark in yours. Because frameworks can be adapted, they matter more than specific tactics, which might land you in trouble if they don't match your reality.

A senior engineer might need to build coalition support before proposing process changes, while a director could simply mandate them. A new manager from an underrepresented background might need to establish credibility differently than someone whose authority is automatically assumed.

The goal isn't about accepting limitations, but rather about working with reality. Sometimes the most effective approach is indirect: finding allies, demonstrating results at a small scale, or reframing proposals to align with existing organizational priorities. Sometimes it's a direct challenge. The skill is knowing which approach fits your context, not ours.

Rapid Assessment When You're New

When you're new to an organization, you lack the historical context that makes diagnosis easier. You can't distinguish between temporary stress and chronic dysfunction, or between individual quirks and systemic patterns.

Start with observation before action. For example, attend meetings without immediately trying to fix them, read Slack channels to understand communication patterns, or follow a deploy or incident response to see how the team actually operates under pressure.

Ask diagnostic questions like the following, which will help you to understand the dynamics rather than just facts:

- What's working well that I should be careful not to break?
- What keeps falling through the cracks despite good intentions?
- Where does the team feel most confident about our current approach, and where do they feel the least confident?

Find multiple perspectives. Quiet team members often have different insights than vocal ones. Similarly, individual conversations reveal information that never surfaces in group settings. Look for patterns across conversations rather than taking any single viewpoint as the complete truth.

In addition, it can be helpful to test small hypotheses. Rather than making significant changes based on incomplete information, run small experiments that reveal how the system responds. Propose a minor process adjustment and observe the resistance or support that emerges.

When Diagnosis Reveals Hard Truths

Sometimes, context assessment reveals that the real problems lie beyond your level or outside your control. The team isn't dysfunctional; they're responding rationally to irrational constraints. The underlying issues aren't just about code quality; they stem from years of business pressure to deliver quickly rather than invest in long-term stability.

In these situations, naming the context becomes crucial. You can't solve every systemic problem, but you can help your team understand what they're navigating and make informed choices about where to invest energy.

A team struggling with constant interruptions might need better boundaries, but they also need leadership acknowledgment that the current system creates unsustainable demands. Similarly, a group frustrated by slow decision making might benefit from clearer escalation paths, but they also need recognition that the broader organization rewards consensus over speed.

In Practice: April's Story

April stared at the incident report from last night's outage: 30 minutes of downtime, angry customers, and her team scrambling to fix a deployment that should have been routine. The system lens revealed to her that they were in chaos during scale-up, with growth outpacing their deployment infrastructure. The human lens revealed her team was exhausted from six months of firefighting. The work lens revealed a previously hidden operational burden that was consuming 40% of their capacity. The cultural lens revealed that PixelCurl prioritized fast shipping over safe shipping.

Every lens pointed to the same conclusion: this wasn't a technical problem that needed a technical solution. It was a systemic mismatch between their infrastructure maturity and business velocity expectations. April realized she'd been trying to solve an organizational problem with better monitoring tools when what she really needed was to shift how leadership thought about platform investment versus feature speed.

Connecting Diagnosis to Action

Context diagnosis only matters if it changes what you do. Each of the following insight lenses should inform specific leadership adjustments rather than just giving you more things to think about:

System insights guide your pace and communication style.

In high-pressure environments, shorter cycles and more frequent check-ins might be necessary. In stable environments, you have more time for consultation and building consensus.

Human insights shape your emotional approach.

Exhausted teams need encouragement and protection. Anxious teams need clarity and predictability. Confident teams can handle more ambiguity and challenge.

Work insights drive operational changes.

Hidden work needs visibility tools. Misaligned work needs priority clarification. Unsustainable work needs capacity protection.

Cultural insights inform your political strategy.

Some environments reward direct challenge, while others require careful coalition building. Additionally, some environments prize individual heroics, while others value collaborative achievement.

The goal isn't a perfect diagnosis; it's a good enough assessment to avoid significant mismatches between what you do and what the situation requires.

In Practice: April's Story

April made this discovery when she realized her team's lack of strategy alignment was a major problem. The system pressures (scale-up chaos), human factors (uneven morale and trust), work patterns (hidden operational tasks), and cultural dynamics (speed rewarded over stability) all required coordinated responses. Her breakthrough came from addressing multiple dimensions simultaneously rather than treating each as a separate problem.

Context diagnosis is a skill that improves with practice. The more situations you navigate, the faster you can recognize patterns and adjust accordingly. But it starts with the discipline to pause and assess before you act, even when the pressure to do something—anything—feels overwhelming.

Thrive in Chaos

When things fall apart, most leaders grab for control. They often think they need to add approval layers or implement more checkpoints. But what results is an environment where everyone feels busy but achieves nothing. The instinct makes sense because chaos feels scary and control feels safe. But what we've learned, after watching teams implode and rebuild, is that control is often the problem, not the solution.

It's the management equivalent of putting duct tape on a cracked foundation. It might hold for a week, but the house is still falling.

The Trap of Bureaucratic Control

Bureaucratic, or hierarchical, control seems attractive when your team is drowning. If deadlines are slipping, trust is shot, and nobody knows what anybody else is doing, it's easy to feel out of control. Leaping to introduce structures is a method that many leaders follow because they think it will make them feel more in control.

So, what do most leaders do? They call more meetings, add more sign-offs, and create more reports that nobody reads but which everyone is still required to write. Now your team is drowning *and* filling out paperwork.

This isn't about being anti-process. Smart teams need good architectural decisions, incident protocols, and clear ownership when things break. But there's a difference between infrastructure that helps people do their job and bureaucracy that becomes the job.

Bureaucratic control feels productive because it creates visible activity—meetings are scheduled, documents are reviewed, and approvals are tracked. But none of that creates forward momentum. When teams are already struggling with broken trust or unclear priorities, adding process overhead deepens the dysfunction rather than solving it.

The Punk Alternative: Motion Creates Clarity

Punk understood this distinction in the 1970s. When the music industry became bloated with gatekeepers and elaborate approval processes, punk didn't try to reform the system; it bypassed it entirely.

Record executives in suits were deciding what people wanted to hear and creating elaborate processes for everything. Want to make music? Get in line behind the gatekeepers.

But punk said, "Forget that approach entirely."

The Ramones, for example, recorded their debut album in seven days. Most bands today spend longer than that picking snare sounds. Minor Threat put out eight songs in under 10 minutes and called it an album. Black Flag lived in a van and played anywhere that would have them—basements, community centers, somebody's garage.

Here's what they understood: motion creates clarity. You don't get good by planning to get good; you get good by doing the thing, failing fast, and doing it again better. The Ramones didn't wait for the perfect studio or the right producer; they just plugged in and played.

As Gorilla Biscuits said, "Let's start today."

That energy—that bias toward action—is what gets teams unstuck when everything else has failed. However, energy alone isn't enough.

From Energy to Discipline

Some punk scenes burned out fast—all energy, no structure. People got exhausted, and the movement collapsed.

But hardcore punk learned from those mistakes. They adopted the same speed and urgency, but with something punk didn't always have: discipline and sustainable practices.

Hardcore bands shared vans and gear. They created zines to keep scenes connected across cities, building networks that lasted because they weren't just about individual expression; they were about community. Minor Threat's "Straight Edge" was more than just a song about not drinking—it was a commitment to showing up clearheaded and focused, to building something that mattered.

That's the evolution every leader needs to make—from creating motion to creating sustainable motion, from moving fast to moving fast together. The music was urgent, but the community was built to last.

Open Source: Punk Principles at Scale

The same ethos that drove hardcore scenes shaped early computing culture, though nobody called it punk leadership at the time.

When something broke, you fixed it yourself. When you needed a tool that didn't exist, you built it. When you figured something out, you shared it so the next person didn't have to start from scratch. It was a deliberate approach to solving problems without waiting for permission.

Linux wasn't built by committees. PostgreSQL wasn't designed by focus groups. Apache didn't come from a corporate roadmap. They came from people who saw problems and solved them, who shared code because making everyone else's life easier made their own life easier too.

Leadership in these communities came from contribution, not titles. Influence was earned commit by commit. Coordination replaced control, and shared values replaced rigid hierarchy.

However, open source also reveals the limitations of purely do-it-yourself approaches. Projects struggle with maintainer burnout, governance gets messy when egos collide, and funding models break down. The lesson isn't that these principles are perfect; it's that they're powerful when you apply them thoughtfully.

What This Means for Seasoned Leaders

You're not starting a band or maintaining a GitHub repository. You're leading engineering teams with real deadlines, actual stakeholders, and quarterly objectives that matter. But the principles still apply, especially when standard approaches aren't working.

When your roadmap doesn't survive contact with reality, don't add more planning meetings. Rather, ship something small and learn from it. When executives are stuck debating strategy, prototype the most likely path and give them something concrete to react to. When everyone knows a process is broken but nobody wants to own fixing it, write the first draft yourself. These aren't heroic gestures. They're practical moves that restore momentum when teams get stuck in analysis paralysis or bureaucratic quicksand.

The key is reading the situation correctly. Some teams need more structure, not less. Some environments punish anyone who circumvents established channels. Moving fast in a compliance-heavy industry can create legal problems that make your speed irrelevant.

The most effective framework is to distinguish between constraints that serve the work and constraints that serve themselves. Then act accordingly.

The Political Reality

Leading like this inside corporate environments comes with career risk, and that's something that experienced leaders need to acknowledge honestly.

Some organizations love unorthodox leaders who get things done, while others punish anyone who colors outside the lines. As a staff engineer or middle manager, you might see exactly what needs fixing but lack the political capital to fix it. This is where the hardcore lesson becomes crucial: sustainable change requires community, not just individual action. Sometimes the right move is starting small, demonstrating results, and letting success create permission for bigger changes. Sometimes it's finding allies who can support you while you experiment with new approaches.

Build the coalition before you build the solution. The goal is to distinguish between constraints that serve the work and constraints that serve the constraint.

When Momentum Matters Most

Punk leadership shows up in specific moments that experienced leaders recognize: when circular debates need decisive action, when teams feel stuck due to unclear priorities, when broken processes consume more energy than they create.

In these situations, the worst move isn't making the wrong decision; it's not deciding at all.

Progress is possible even when conditions aren't perfect. Sometimes the best way forward is the simplest: stop waiting for permission and start fixing what's broken.

As Youth of Today put it, "It's time to make a change." In chaos, trying with intention and discipline beats having the perfect plan every time.

Focus on Outcomes

In unstable environments, teams grab onto what they can measure, such as sprint velocity, story points, and deployment frequency. The standups are full, the roadmap is moving, and code gets shipped. From a distance, everything looks productive. But when you ask, "What exactly did we achieve?" the answers become vague.

This isn't laziness or lack of focus; it's a survival response. When everything feels uncertain, measurable activity provides psychological safety. But activity without direction burns people out faster than honest idleness. Don't confuse activity with progress.

Your job as a leader is to cut through the motion and point toward impact that actually matters.

The Sophistication Trap

Experienced teams fall into activity traps differently than novice ones. They don't track meaningless vanity metrics or celebrate meaningless milestones. Instead, they prioritize work that seems sophisticated and feels important, even if it disconnects from business reality: architecture discussions that never conclude, refactoring sprints that improve code quality but don't unlock new capabilities, elaborate planning processes that consume more time than the work they organize, and process improvements that make work feel more organized without making it more effective.

While none of this is wasteful in isolation, these activities become traps when they turn into ends in themselves rather than means to customer or business outcomes. Teams can spend months perfecting systems that don't solve problems anyone actually has.

Why Outcome Frameworks Miss the Point

Most outcome frameworks, such as objectives and key results (OKRs), fail because they address symptoms rather than causes. Teams write better objectives, track cleaner metrics, and still drift toward irrelevance. The problem isn't the measurement technique; it's the assumption that clarity comes from better goal setting rather than better context understanding.

Outcome focus isn't about writing perfect OKRs or creating elaborate dashboards. It's about building the organizational muscle to separate meaningful progress from mere activity, to recognize when work matters and when it doesn't, and to make trade-offs based on impact rather than ease or familiarity.

This requires different approaches in different contexts. A startup team needs to iterate quickly toward product–market fit, accepting that many experiments will fail. An enterprise platform team needs to make infrastructure investments that might not show business impact for quarters. A team inheriting legacy technical debt needs to balance maintenance with feature development.

The framework isn't the same metrics for everyone, but it is the same discipline—connecting daily work to meaningful outcomes, even when those connections are indirect or delayed.

When Outcomes Are Genuinely Unclear

Real leadership situations don't come with predefined success criteria. You're often working through genuine uncertainty about what outcomes to pursue, let alone how to measure them.

Exploratory product work doesn't have predetermined outcomes. You're iterating toward understanding, not executing against a known goal. Success might mean discovering that an approach doesn't work, which prevents months of wasted effort.

Technical architecture decisions have indirect business impact. The outcome might be enabling future development speed or reducing operational risk, neither of which shows up in quarterly metrics. You're building capabilities that enable other people to create value.

Platform and infrastructure investments create value through what they enable rather than what they directly produce. The outcome is often invisible to end users but crucial for engineering productivity. You're optimizing for leverage, not immediate output.

In these situations, outcome focus means being explicit about the hypothesis you're testing or the capability you're building, even when the business impact is uncertain or delayed. The goal is to arrive at an honest acknowledgment of what you're trying to achieve and why it matters.

The Political Dimension

Outcome clarity becomes complicated when your team's success conflicts with other stakeholders' incentives. Your engineering team might optimize for system reliability while the business team optimizes for feature velocity. Your infrastructure improvements might slow down feature delivery in the short term. Rather than being technical problems to solve with better metrics, they're political realities to navigate with more transparent communication and stronger alliances.

Sometimes the right move is reframing your outcomes in language that aligns with broader organizational priorities. Technical debt reduction becomes "increasing feature delivery predictability." Infrastructure investment becomes "reducing business risk from outages." You're translating engineering value into business language.

Sometimes you need to surface the tension and help leadership make informed trade-offs explicitly. For example, you could say, "We can ship faster by accumulating technical debt, or invest in stability and move more deliberately. Both have costs." The sophistication is knowing when to align, when to push back, and when to document disagreement while moving forward.

Context Shapes Outcome Focus

Your approach to outcomes depends heavily on your diagnostic assessment. The same outcome might require completely different approaches based on your team's situation. Let's consider the four contexts we've been discussing.

System context determines your time horizon. Startup environments might measure outcomes in weeks. Enterprise environments, on the other hand, plan for quarters or years. Platform teams might work toward capabilities that won't show impact until other teams build on them.

The human context will affect how you communicate outcomes. Burned-out teams need to see progress quickly to rebuild confidence, while energized teams can sustain longer feedback loops if they understand the strategic rationale. Teams with low trust need smaller, more frequent wins.

Work context reveals whether outcome clarity is genuinely the problem. Teams drowning in hidden support work don't need better goal setting; they need workload visibility and protection. Teams with clear outcomes but broken processes need operational fixes, not strategic realignment.

Finally, the cultural context shapes what outcomes are politically viable. Some organizations reward visible feature delivery over infrastructure investment, but others punish anything that looks like it's moving too fast. You need to work within these constraints while gradually shifting them.

Failure Modes of Outcome Obsession

Outcome focus can become counterproductive when taken to extremes. This section outlines some of the warning signs to look out for that signify you've crossed the line.

Measurement illusion

This could include spending more time on tracking progress than making it, creating elaborate dashboards that nobody uses for decision making, or optimizing for metrics that don't actually correlate with value.

We'll talk more about metrics in Chapter 10.

Outcome rigidity

This includes refusing to adapt when you discover better approaches or when context changes, or treating initial outcome statements as unchangeable contracts rather than working hypotheses.

False precision

This refers to pretending that you can measure things that are genuinely unmeasurable, like team morale or code quality, rather than using qualitative assessment and human judgment.

Gaming

In these cases, teams are optimizing for metrics rather than the underlying value and improving deployment frequency by breaking large changes into smaller, meaningless commits—hitting feature delivery targets by cutting quality or maintainability.

The goal is outcome awareness, not outcome obsession. You want clarity on what matters without sacrificing the flexibility to adapt as you learn more.

What This Looks Like in Practice

Consider a product team that is building a new feature. Instead of measuring completion percentage, you can track early user engagement signals. When engagement is lower than expected, you can pivot the approach rather than pushing toward the original deadline.

As another example, an infrastructure team might be migrating to a new deployment system. Instead of measuring migration percentage, you can track time from code to production and assess the rollback success rate. When the new system performs worse than the old one, you can pause to fix core issues rather than continuing the migration.

Now consider a platform team that is building internal developer tools. Instead of measuring adoption rates, you can track developer productivity signals like time to onboard new team members or frequency of deployment pipeline failures. When productivity doesn't improve, you can investigate workflow bottlenecks rather than adding more features.

Each example shows the same pattern: define outcomes that connect to larger value, measure leading indicators rather than lagging ones, and maintain a willingness to adapt when reality conflicts with the plan.

Making It Sustainable

Outcome focus requires organizational discipline, not just individual discipline. You need systems that reinforce good decision making rather than hoping people will remember to think about impact. This includes:

- Regular outcome reviews that ask "What did we achieve?" rather than "What did we ship?"
- Team retrospectives that examine whether completed work actually solved problems
- Planning sessions that start with desired outcomes rather than available work

The goal isn't perfection but direction. You want teams that can answer "Why does this matter?" for the work they're doing and "How will we know it worked?" for the changes they're making.

When outcome focus becomes habitual rather than forced, teams waste less energy on work that doesn't matter and invest more energy in work that does. That's the difference between being busy and being effective.

Shift Between Roles

Different moments call for different leadership styles.

Most leaders fail not from lack of skill but from rigid application of what worked before. They've mastered one approach—coaching, directing, or consensus building—and apply it regardless of context. That approach works well in stable environments where situations repeat predictably. However, in chaos, it's a liability.

The hardcore punk scene understood this intuitively. Different moments demanded different energy and approaches. Sometimes you needed strategic thinking to see the bigger picture. Sometimes you needed to tend to people who were struggling. Sometimes you had to get your hands dirty and do the work yourself.

The effective leaders could read the room and adjust accordingly, expanding their range so they could meet the moment with what it actually needed.

The Cost of Leadership Rigidity

Engineering leaders get trapped in familiar patterns for understandable reasons: a vice president who has built credibility through technical depth continues to delve into implementation details; a director who has earned trust through careful consensus building continues to facilitate when decisions need to be made quickly; a staff engineer who influences through expertise struggles to step back when the team needs space to learn.

These aren't character flaws; they're successful patterns that become limitations when overused. When the leader doesn't recognize when the context has shifted, their default approach might no longer serve the situation.

The problem deepens when your identity becomes fused with your style. "I'm a service-oriented leader" or "I lead through vision" stops being a description and becomes a constraint. You start defending your approach rather than questioning whether it fits the moment.

But leadership isn't a personality trait or a brand you maintain—it's a situational skill that adapts to what the team needs right now.

Three Essential Modes

You don't need to excel at every leadership style, but you need the ability to consciously shift between a few key modes based on what the situation demands. The three roles discussed in this section cover most of the territory you'll face in chaotic environments.

The pilot: Zooming out for direction

Being the pilot means holding a broader view. These leaders connect work to purpose, surface competing priorities, and help teams navigate when the path isn't clear.

This isn't about having all the answers or being the smartest person in the room, but about creating clarity when everyone else is stuck in the details. You zoom out to see patterns, dependencies, and trade-offs that aren't visible at ground level.

Use pilot mode when priorities are unclear, teams are stuck in tactical debates, or you need to make sense of competing pressures from different stakeholders.

This could look like:

- Stepping into a circular architecture discussion and reframing around user impact
- Helping a team see how their technical decisions connect to business objectives
- Surfacing the real trade-offs when everyone's talking past each other

When it backfires. Vision-setting can feel disconnected if the team is overwhelmed by immediate operational problems. Additionally, some cultures resist top-down direction, even when it's collaborative. Your credibility for strategic thinking may depend on your background and position in the organization.

The medic: Stabilizing and rebuilding trust

The medic role focuses on people and relationships. They listen without agenda, help process difficult situations, and create space for recovery when things have been intense.

Rather than being a type of therapy or performance management, it involves recognizing when emotional dynamics are blocking progress and addressing them directly, instead of hoping they'll resolve on their own.

Use medic mode when tension is rising, someone's burned out or isolated, or when a difficult decision has landed and people need space to process.

This could look like:

- Pulling aside a senior engineer who's been unusually quiet and discovering they're overwhelmed by scope changes
- Facilitating a conversation between team members who've been talking past each other
- Acknowledging that a reorganization was disruptive rather than pretending everyone should just adapt

When it backfires. Emotional labor can become an expectation rather than a choice, especially for leaders from underrepresented backgrounds. In addition, some organizational cultures punish leaders who spend time on "soft" issues. Your capacity for this work has limits that need protection.

The engineer: Getting into the work

The engineer role moves from observation to participation. They pair on difficult problems, debug blocked processes, and contribute directly rather than just directing from above.

Instead of being considered micromanagement, think of engineering mode as recognizing when the team needs you to roll up your sleeves and work alongside them, especially when technical complexity or organizational friction is creating barriers.

Use engineer mode when momentum has stalled despite clarity, when critical details are blocking progress, or when the team needs to see that leadership is with them, not just watching from above.

This could look like:

- Pairing with a junior developer on a complex migration
- Sitting in on customer calls to understand requirements that keep changing
- Writing the first draft of a process document rather than asking someone else to figure it out

When it backfires. Hands-on involvement can undermine team autonomy if overused. In addition, your technical skills may be outdated compared to those of your team, and getting pulled into implementation can distract from strategic responsibilities that only you can handle.

The Switching Skill

The real capability isn't mastering any single mode but developing the judgment to switch consciously rather than subconsciously. This requires several key abilities.

Pattern recognition

Certain signals will tell you that the current approach isn't working: teams talking past each other, energy dropping in meetings, work stalling despite apparent activity, and learning to read these signals quickly and accurately.

Mode flexibility

Can you shift approaches without losing authenticity? Some roles will feel more natural than others based on your background and strengths, but effectiveness matters more than comfort.

Switching costs awareness

Frequent role changes can confuse teams who need some predictability in leadership approach. The goal is conscious adaptation, not constant shifting. Teams need to understand why you're changing approaches, not just experience the change.

The Reality of Constraints

It's important to be aware that your ability to switch roles is limited. Identity, organizational position, and cultural context all affect which modes are available to you and how they're received.

The goal isn't to be everything to everyone, but to expand your effective range within your actual context and constraints.

Leading Through Others

As you move into more senior roles, your mode switching becomes more complex. You're not just choosing your own approach but coordinating with other leaders and modeling flexibility for your team.

This means helping your direct reports develop their own range rather than expecting them to mirror your style. A director who can shift between pilot and medic modes might work with managers who each have different strengths, creating collective range rather than individual completeness.

It also means reading the broader organizational context and helping your team understand when to match the moment versus when to maintain consistency. Sometimes the business needs everyone aligned on execution, and sometimes it needs different teams operating in different modes simultaneously.

The skill is knowing when your switch should be visible to the team and when it should be invisible. Sometimes you want people to understand why you're changing approaches. Sometimes you just want them to experience the right leadership for the moment.

What This Looks Like in Practice

Consider the following examples of what this looks like in practice:

Pilot mode
> Your team is debating technical architecture but losing sight of user impact. You shift into pilot mode, step back from the technical details, and help them connect their decisions to customer outcomes.

Medic mode
> A senior engineer has been unusually withdrawn since the last reorganization. You shift into medic mode, create space for a private conversation, and discover they're questioning whether their expertise is still valued.

Engineer mode
> A critical migration is stalled because of complex integration challenges. You shift into engineer mode, sit with the team, and work through the technical details until the path forward becomes clear.

The switching happens in response to what the team needs, not what feels comfortable or familiar. Over time, mode switching becomes habitual rather than forced, and your team learns to expect adaptability rather than rigid consistency.

That's the evolution from good technical leadership to effective organizational leadership, where you meet the needs of others rather than providing what you prefer.

In Practice: April's Story

April walked into the post-incident meeting and immediately read the room. The business team wanted answers and timelines. Her senior engineers looked defensive and frustrated. The junior engineers were quiet, probably blaming themselves. April realized that the three different situations would require different approaches within the same meeting.

She started in pilot mode, connecting the outage to broader infrastructure needs and business impact. When she saw tension building between her team and product leadership, she switched to medic mode, acknowledging the stress everyone felt and creating space for honest conversation about capacity constraints. Finally, when the discussion stalled on technical details, she shifted to engineer mode, diving into the deployment process with her team to map out specific fixes. By the end of the meeting, they had achieved alignment on both immediate fixes and longer-term platform investment, moving beyond blame and defensiveness.

Exercise: Practice Your Switches

This chapter is about developing range: the ability to read chaos and respond with what the situation actually needs. This exercise will help you to practice the four leadership switches using a real situation you're facing.

Setup: Pick Your Chaos

Think of a current situation where standard leadership approaches aren't working—something where you've tried the obvious solutions, but progress has stalled. Write it down in one sentence.

Switch 1: Diagnose Context

Run your situation through the four lenses. Be honest; this only works if you see what's actually happening:

System lens
> Is this early-stage chaos (existential pressure), scale-up chaos (growth outpacing systems), or enterprise chaos (politics and interdependencies)?

Human lens
> Are your people energized and ready for challenge, or exhausted and needing stability first? How's the trust level?

Work lens
> Is the real problem hidden work, unclear priorities, or unsustainable firefighting?

Cultural lens
> What leadership approaches get rewarded versus punished in your environment? What constraints do you actually face?

Write down the biggest insight from this diagnosis.

Switch 2: Thrive in Chaos

Based on your diagnosis, what does this moment need?

More momentum
> What's the smallest thing you could ship, fix, or decide this week to break the current pattern?

More control
> What structure would actually help people do their job (not just make you feel better)?

Different focus
> What bureaucratic illusion could you eliminate to free up energy for real work?

Switch 3: Focus on Outcomes

What would actually happen if you solved this?

Don't write about "better communication" or "improved processes." Write about the business or technical outcome that would make someone's life genuinely better.

Switch 4: Shift Between Roles

Given everything above, which role does this situation need right now?

Pilot
> Should you zoom out and help people see connections they're missing?

Medic
> Does someone need space to process before you can move forward?

Engineer
> Do you need to roll up your sleeves and work alongside the team?

Write down which mode you'll try first and one specific action you'll take in the next three days.

Reality Check

Look at what you wrote. Does it match what you've been doing, or does it suggest a different approach? If you're not planning to change anything, run through the diagnosis again. You probably missed something important.

The point isn't to solve everything perfectly. It's to practice reading chaos and choosing conscious responses instead of defaulting to what feels familiar.

Conclusion: Pulling It All Together

Engineering leadership rarely starts in calm waters. You inherit chaos in all its glory. You don't get to wait for someone else to fix it. That's the work.

What we've learned from punk, hardcore, and open source is that you don't beat chaos with control. You don't earn trust with posturing. You don't build momentum by waiting for the whole picture. Instead, you act, commit, and teach. Show up again and again, because that consistency is what turns noise into clarity.

The four switches we've covered aren't separate tools; they work together.

Context diagnosis helps you determine which moments require which approach. *Thriving in chaos* creates momentum that aligns with what you've read. *Outcome focus* keeps you pointed toward impact as you adapt. *Role flexibility* gives you the range to meet different situations with what they need.

Not every approach survives and not every team makes it through. But the spirit and the ethos of building, adapting, and persisting outlasts the moment and creates something bigger than any individual effort.

This is how you turn chaos into productivity. This is how you lead when the map doesn't match the territory and the rules keep changing. You read the context, focus on what matters, switch between what the team needs, identify the hard choices, and build systems that enable others to do the same.

Remember: range beats rigidity and motion creates clarity and persistence, with principles that outlast any single crisis. That's leadership in chaos—and how you build something that lasts.

In Chapter 4, we'll talk about building cohesive teams using some of these switches.

Building Cohesive Teams

In Chapter 3, we acknowledged that you likely didn't get this job because things were going smoothly and that, more often than not, engineering leaders step in when something is broken. And nothing is better at breaking teams than chaos.

When Juan joined Splice as its first engineering leader, the team was about to raise a major round of funding and was preparing for explosive growth. From the outside, it looked like liftoff. Inside, it was barely held together: 12 people—5 of which were engineers—fragile engineering fundamentals, and no real team structure. It was just a team of smart but exhausted individuals sprinting toward burnout. That's not unusual; most leaders don't inherit a cohesive team. Often, they inherit chaos, drift, and the pressure to fix it—fast.

You don't fix that with process. You fix it by rebuilding the team from the inside out. That starts with psychological safety, followed by functionality, and then capabilities. Only once those foundations are in place should you grow.

This chapter is your roadmap for that journey. And if you're not stepping into a new team right now, pretend you are. Look at your current team with fresh eyes. What would you see if you'd just inherited it? Where would you start?

It Starts with Safety

You walk into the room, and suddenly everything's a little quieter. People hold their questions a beat longer. They start editing themselves. Everyone's wondering the same thing: "What does this new leader mean for me?"

That's not insecurity; it's a kind of survival instinct. Change can feel threatening, especially for teams that are overworked, have had negative past experiences, or lack information. You can't drive change if your team feels under threat.

If you want to fix or improve anything, your first job is to lower the threat level. In case you haven't read it, Google's *Project Aristotle*,[1] the company's massive study of what makes teams effective, found that psychological safety was the number-one factor. It wasn't intelligence. It wasn't tenure. It wasn't tooling. It was safety—the ability to speak up without fear. If your team doesn't feel safe being honest, you won't get genuine feedback. You'll just get silence.

In Practice: April's Story

When April joined her first PixelCurl team meetings, she noticed the pattern immediately: cameras off, short answers, no one jumping in to help when someone struggled to explain a blocker. She asked, "What have you been struggling with lately?" The chat stayed silent. Finally, a senior engineer unmuted: "Nothing that we can't handle." April recognized the self-protection. At the next meeting, she tried something different. She shared her screen showing her chaotic attempt to understand their deployment pipeline and said, "I've been here two weeks, and I still can't figure this out. I watched the recording from last week and got completely lost." In chat, a junior engineer typed, "lol same—Marcus and Jordan both do it differently and there are no docs." The floodgates opened. People started sharing frustrations in chat and voice. April didn't solve anything that day, but she lowered the threat level enough that people felt safe being honest, even over a video call.

To be clear, safety doesn't mean comfort, and it doesn't mean coddling. It means creating the conditions where people can operate with honesty, courage, and clarity, without fear of punishment or humiliation. In practice, this looks like the following:

- Showing up curious and calm
- Protecting your team publicly
- Naming the tension, and acknowledging any uncertainties or fears
- Normalizing learning and failure
- Listening more than you speak (at least at first)

If you hold yourself to these conditions as a foundation, you stand a good chance of being able to build an effective and cohesive team.

---

1 Google's *Project Aristotle* identified psychological safety as the most important factor in team effectiveness. The full findings are available at *rework.withgoogle.com*.

Get a Functional Team, Fast

When you take over a struggling team, the problems are rarely subtle. People are juggling too much, and priorities are shifting weekly. Deadlines are slipping—or worse, being met at the expense of work–life balance and quality. No one's confident, and everyone's tired.

In these moments, there's a temptation to go big—to set a bold new direction, roll out a better process, or fix the culture. But trying to solve everything at once almost always makes things worse.

So, what should your first milestone be? Get the team doing good work. It doesn't have to be perfect work, just good enough to keep moving forward. Focus first on stabilization, not optimization. Your goal is to create enough clarity and space for progress to happen without causing new problems. This helps the team get a win and take a breather, while also giving you room to think through the next steps.

Functionality is the goal; cohesion and performance come later.

Understand the Work

Before you start fixing anything, we encourage you to review the work ahead of the team. Yes, you should talk to your crew and others around the organization, but those accounts will come through a human filter. A feature is a feature. The implementation may vary, but the outcomes should be clear. So, dig into the work—not what's on the roadmap, not what the strategy doc says, and not what you think the team *should* be doing. Focus on what they're *actually* doing, day to day, task by task, under real-world conditions.

That means observing the work closely. For example, you could sit in on standup meetings, watch a deploy, or ask someone to walk you through their week. Look at what's in progress, what keeps slipping, and what's quietly draining energy. We've found that understanding the work is one of the most reliable ways to get a clear picture of a team's state.

If the work seems viable, well scoped, documented, reasonably planned, and generally moving forward, then the next place to look is the team. Are they equipped? Are they skilled and cohesive? Are there dynamics slowing them down? If the work itself is unclear—poorly scoped, with unrealistic expectations or constant pivots—then the problem may lie elsewhere: in planning, dependencies, cross-team agreements, or upstream leadership.

When a team isn't performing well, there's usually some misalignment at the core. And there are always two sides to the story. Consider "The team missed a deadline," versus "We were given a date without a chance to agree to it."

Understanding *why* things are going wrong is often more important than identifying *what* is going wrong. A missed ship date isn't the diagnosis; it's the symptom. The root causes could include capacity issues, skill gaps, planning debt, weak leadership, or a combination of all of these.

Start by making the problems visible. Follow the work, and let it show you what's actually happening. From there, you can begin mapping patterns and guiding the team with clarity.

In Practice: April's Story

April spent her first week at PixelCurl doing precisely this. She reviewed every ticket in flight, attended standup meetings with minimal participation, and watched a complete deployment cycle from planning to production. What she discovered didn't match what she'd been told in interviews. The roadmap showed a clear focus on new rendering features, but the actual work told a different story: 60% of engineering time was going to support requests from the sales team for custom client integrations. Instead of being tracked as projects, they were handled as "quick favors" that often took weeks. The team was exhausted, not due to a lack of skill, but because they were tasked with building a custom product for each new customer while also pretending to deliver a standard platform. The work made it clear—this wasn't a team problem; it was a strategy problem that no amount of process improvement could fix.

Renegotiate Commitments

You will inherit promises you didn't make. Some of them are buried in backlogs. Some are recorded on a slide from a board meeting six weeks ago. Some were made in good faith under bad conditions, by people who are no longer around.

And now those promises are the team's problem. This is where a lot of new leaders slip. They want to prove they can execute and be seen as dependable, so they protect commitments at the expense of the team. Do not do that.

There's a window, right when you take on a new team, where you have permission to make bold decisions. You can ask hard questions, call out broken plans, or pause work or plans. This is trust credit. You haven't delivered yet, but people are hoping you might be the one who can turn things around.

Use that window well. It won't last forever, but it can buy you the time to stabilize and rebuild. Removing unrealistic milestones might feel like pushing back, but done right, it builds trust. When you advocate for a plan that the team believes in, they start to believe in *you*. You're not just protecting them from bad decisions; you're giving them a chance to succeed. That matters.

But be careful not to push it just to prove a point. This isn't about being right; it's about being effective. If you want to renegotiate, then be convincing. Come with alternatives and data. Present a clear-eyed assessment of risk.

In some ways, influence matters more than certainty. However, influence is built on understanding reality. If you want to lead effectively, you must first diagnose where your team actually is before determining where they can realistically go.

Resetting expectations doesn't mean lowering ambition—it means protecting focus and giving the team room to deliver sustainable progress without burning out or cutting corners on quality. This requires an honest assessment of current capacity constraints and what is genuinely achievable given the team's context.

Do this early, before delivery pressure mounts and before morale erodes. When you wait until people are already struggling to hit impossible targets, renegotiation feels like failure rather than smart leadership.

Ask yourself the following questions:

- Who made this commitment?
- Was the team involved?
- What tools and timeline would we actually need to deliver this?
- What is the cost of struggling quietly versus renegotiating openly?

This is one of your earliest leadership tests: will you uphold the illusion of progress, or create the conditions for real momentum?

Remove Blockers

If you want to build trust quickly, then you need to clear a path.

Most teams aren't blocked by big, strategic challenges. They're blocked by 100 small things that no one's had the time or authority to fix—for example, a broken staging environment, a flaky deploy, or an approval process that slows everything down. These aren't side issues; they're morale killers.

When you join a new team, one of the best ways to earn credibility is to ask, "What's slowing you down?" And then, do something about it. That's leadership.

You don't need to fix everything yourself. However, you do need to demonstrate that you're paying attention and understand the cost of friction. Show that you care about what's making the work harder than it needs to be. Start with small wins:

- Identify something that frustrates the team and resolve it within a week (or a day).
- Remove a painful dependency.

- Cancel a pointless meeting.
- Rewrite a tool permission that's been wrong for months.
- Purchase an inexpensive (e.g., $20/month) tool that the team has needed for a while.

It's not about the size of the fix; it's about the message it sends. Removing blockers shows that you're working with the team, and not just observing them. And more importantly, you're willing to spend your influence on things that make their work easier. This is how you start shifting the energy and improving the team—not with grand speeches, but with tests that work, resolved issues, and a smoother path forward.

Get Capacity Under Control

Most new leaders look at a roadmap and think they're seeing the work that needs to be done. But capacity isn't just what's in the task tracker; it's everything that's happening behind the scenes. And if you don't get close to it early, you'll mistake disorganization for progress and overload for engagement.

Sometimes, capacity is taken up by untracked tasks:

- A founder asking for a "quick change"
- A product manager slipping in a "small tweak"
- Another urgent issue to resolve
- A detour into a popular new programming language because someone's curious and motivated

None of this is unusual, but it adds up—fast. So, it's important to start with a straightforward question: "Where is the team's time going?" Don't aim to shame anyone, but gather the facts.

Then get that inbound work under control. All work, whether planned or unplanned, internal or external, needs to be tracked and visible. If it's not tracked, it can't be prioritized. And if it can't be prioritized, the team will focus on whatever seems most urgent or interesting.

Set a basic standard:

- All work is visible.
- All work is prioritized.
- We only work on what matters.

And help the team enforce it. Saying no to a founder isn't easy. Redirecting a well-meaning leader asking for "just five minutes" takes confidence and clarity. So don't leave that burden on individual contributors. Give them clear language and the permission to do so. Be proactive, and send a short note to other leaders: "To help the team stay focused and deliver better, we're routing all new work requests through a shared intake process. This will help us make better trade-offs and protect capacity for what matters most."

Then follow through by reinforcing the message and repeating it when people forget. This behavior won't change overnight, but your consistency will shape expectations.

In Practice: April's Story

April implemented this at PixelCurl immediately. The StudioOps team was overwhelmed by interruptions: founders messaging engineers directly on Slack about demo features, product managers reassigning sprint work midweek for "urgent" client requests, and constant context switching. April sent a brief note to leadership, routing all new requests through sprint planning, and then backed it up. When a founder messaged an engineer directly, she gently redirected: "Added this to our intake board for Monday planning. If it's blocking you before then, let me know and we can discuss trade-offs." The first few weeks were uncomfortable, but within a month, the team's velocity increased because they could focus on completing tasks without frequent interruptions.

Also, remember that this kind of structure is just a starting point. A rigid intake system might feel restrictive at first, but it's temporary. Over time, the team should own its priorities and trade-offs. You're just giving them the clarity and support to get there. Driven people want to do meaningful work. Your job is to help define what *meaningful* looks like and protect the space to achieve it.

Finally, remember to shift the balance. While every team will have some unplanned, reactive, low-leverage work, if that's all that they're doing, then they're just maintaining operations rather than building.

Understand Skills, Dynamics, and Performance

As everything else is being set, work, flow, and focus start with observing the team. Who does what? Who leads without being asked? Who quietly shoulders more than their share? Who trusts whom? Who avoids whom? Where does energy go, and where does it get stalled?

You're not just here to build a delivery system. You're here to understand the people inside it. Every team has a map. And we're not talking about the org chart. We're talking about the map that is made of trust, influence, capability, and tension. Your job is to learn that map quickly in order to use it effectively.

Talk to everyone: engineers, designers, leads, quiet contributors, recent hires, and people who have been there since the beginning. In these conversations, you're listening for patterns:

- Who fills in the gaps?
- Who never asks for help?
- Who's disengaged?
- Who's stretched too thin to be effective?

The team knows. They've already adapted to what works and what doesn't. Your job is to bring that knowledge to the surface, make sense of it, and act on what matters. And when you see performance issues, don't wait. If someone is struggling, chances are the rest of the team has noticed. They're waiting to see whether you will, too—whether you'll acknowledge it and take action.

In Practice: April's Story

April faced this with Marcus, a brilliant senior engineer at PixelCurl who worked in total isolation. He dismissed code review feedback, refused to pair with juniors because it "slowed him down," and had created a section of the codebase that only he understood. The team had adapted around him for months. April knew everyone was watching. She decided to have a direct conversation with him: "Your technical skills are exceptional, but I need you to pair with junior engineers twice a week and engage constructively in reviews. This isn't optional." Of course, Marcus pushed back and tested her resolve for three weeks. But she held firm. Eventually, he chose to change his ways rather than leave the company, and six months later became one of the team's strongest mentors. The junior engineers who'd been watching learned that April meant what she said about team health.

You don't need a process. You need clarity. Be honest and direct. Set expectations and offer support. Don't let issues linger. Unaddressed issues don't stay quiet; they turn into bigger problems and slow progress. And they signal to the team that outcomes don't matter.

So, be kind and give helpful feedback. Remember: you set the bar. If you allow poor-quality work to slip through, don't be surprised when that becomes the new standard.

Capacity Is Not Enough; Build Capabilities

While getting a team to be functional is a win, it's not the end goal. It just means the issues have been addressed, the work is flowing, and people can breathe again. Now you need to help the team grow, but not by adding more people or repeating the same tasks.

This next phase is about building something deeper: capability. Capabilities enable a team to tackle more complex problems without breaking down. It's what makes progress sustainable and turns a group of people who can deliver into a team that can lead and improve.

You're shifting from "Can we deliver this sprint?" to "Can we consistently deliver value, even under pressure?" You're moving from "Can we execute?" to "Can we improve how we operate, together?" That is what capable teams do.

Capabilities as a Team-Level Tool

This idea was inspired in part by Camille Fournier's post on engineering ladders,[2] a tool that helped bring structure and fairness to how individual engineers grow. But teams don't usually get that same clarity. We expect them to improve without ever defining what "better" looks like.

A capability framework fills that gap. It defines the skills and behaviors that make teams reliable, adaptive, and resilient—how they collaborate, communicate, and respond when challenges arise. Its purpose is guidance, not evaluation. It works as a tool, providing teams with a shared language for what good looks like, a means of self-assessment, and a path for growth together.

Just like engineering ladders, capability frameworks need structure and care to be useful. It's not about creating a checklist; it's about creating clarity.

Here's how we've seen it work:

Define team capabilities.
> Start by naming the essential skills and behaviors that enable effective teamwork. Examples include navigating ambiguity, sharing progress clearly, or managing dependencies. Keep them tight and clear—enough to guide reflection without overwhelming the team.

---

2 Fournier's post offers a framework with structured levels (*https://oreil.ly/e2hzi*) and expectations for individual engineer growth, helping create transparency in career progression within engineering organizations.

Assess team effectiveness.

Let teams evaluate how they're doing against those capabilities. Use structured conversation and input from partners, not just gut feelings. This process is gathering actionable insights, not assigning grades.

Identify areas for improvement.

Most teams won't be strong across the board. That's fine. Use the assessment to focus the team's energy. What's dragging them down? What limits trust, delivery, or learning? Pick a few areas to improve.

Provide a path to grow.

Each capability should have clear examples of success and observable behaviors Provide the team with concrete guidance to aim for. The goal is progress and momentum, not perfection.

High-capability teams don't just work more. They work more effectively, with less conflict, stronger trust, and greater impact.

What Great Teams Do

Enough with frameworks—let's get specific. What exactly do capable teams do differently in practice?

We've identified seven core capabilities that distinguish the teams that deliver from the teams that get stuck. These aren't personality traits or abstract ideals. They are observable behaviors that can be discussed and improved. Not all teams need to excel at everything at once, but high-performing teams eventually build strength across all capabilities.

Use this as a diagnostic tool. Where is your team strong? Where are they struggling? What's the next thing worth focusing on?

They turn ambiguity into action

Most teams are handed vague goals and unclear priorities. The difference is how they respond. Strong teams don't wait around for perfect clarity. They ask focused questions, connect their work to actual customer problems and business outcomes, and propose direction rather than waiting for it to arrive fully formed.

When a project manager says, "We need to improve onboarding," a capable team responds with, "Are we trying to reduce drop-off in the first session, or increase activation by day seven? Those are different problems."

They clarify early, get stakeholder feedback before they're too far down a path, and make sure everyone understands not just what they're building, but why it matters.

A team that can't do this will spin for weeks building something important-ish. A team that can do this will define measurable goals, connect them to the company strategy, and move forward with confidence even when things are uncertain.

They communicate relentlessly (and don't make people chase them)

Good teams share progress consistently. You don't have to chase them down to know where things stand. They share status early and often and surface blockers before blockers become crises. They maintain predictable channels: weekly updates, standups that highlight risk, and Slack threads that document project progress. When they need help, they ask for it with enough context that someone can actually provide support.

Here's the test: can someone outside your team explain what you're working on and why? If not, the team isn't communicating enough.

We've seen teams go silent for three weeks, only to surprise everyone with "actually, we're behind." That's not a communication problem; it's a trust issue. Compare that to teams that raise issues early on in standup meetings: "We hit an unexpected dependency with Platform. Can we renegotiate the timeline? We propose pushing the launch back one sprint." The outcome will be the same, but they show different levels of capability.

The best teams also make the purpose of their work clear—not just to leadership, but to the rest of the company so that people understand why the work matters.

They manage dependencies without drama

No team works in a vacuum. Capable teams know this and plan for it. They track cross-team dependencies proactively. They communicate with other teams directly, early, and clearly. When something breaks down, they collaborate to fix it before escalating. And when they do escalate, they do it to clarify ownership or get an executive decision, not to assign blame or cover their own reputation.

Weak teams get blocked and then shrug: "We can't ship because Team X didn't deliver their API." Strong teams see that coming two weeks out, raise it in planning, and work with Team X to either derisk it or adjust scope. They maintain shared accountability for shared outcomes.

If your team is constantly surprised by dependencies, or if they treat other teams like ticket systems instead of partners, you've got work to do here.

They execute with discipline, not heroics

Capable teams ship predictably—not because they work harder, but because they plan better. They understand their capacity and don't overcommit. They adjust scope and timelines based on what's actually happening, not what they wish were happening.

They refine their process continuously, tweaking standups, improving tooling, and running retrospectives that lead to actual changes.

The goal isn't to ship faster, but more sustainably, at a cadence that's observable and repeatable. When a team says, "We need to double our headcount to hit our commitments," that's usually a sign that they're operating without discipline. A capable team would instead say, "We're paced to 70% load and shipping every two weeks. We're iterating on our retrospectives to reduce rework and improve flow. If priorities expand, let's discuss scope trade-offs."

Execution is about finding a rhythm that works for your specific team.

They see risk coming and know how to respond

Things will go wrong. The question is whether your team is ready when they do. Resilient teams identify risks early. They build monitoring that actually catches problems before customers notice, and they have incident protocols that are both documented well and practiced. When something breaks, they respond quickly and calmly—informing the right people, sending frequent status updates, and keeping stakeholders up-to-date.

After the incident, they run a neutral assessment and actually implement the follow-up items, using incidents as opportunities to improve, not as blame exercises. Let's contrast two scenarios:

Team A
> They find out something is broken only when a customer posts on social media about it. They scramble, and no one is sure who should be responding. Updates trickle out inconsistently.

Team B
> Their monitors flag an issue at 2:00 a.m. The on-call engineer pages the appropriate backup. They follow their runbook and resolve the issue within 40 minutes and post a detailed incident report by the end of the day.

Both teams had an outage, but only one of them was ready for it.

They invest in team health (because it's not optional)

Healthy teams don't just deliver; they sustain themselves. They onboard new people with intention. For example, every new teammate could be assigned a buddy, a 90-day plan, and ample opportunities to ask questions without feeling stupid. Leaders that invest in team health will create working agreements and actually revisit them when the team changes. They support knowledge sharing, such as documentation, pairing, and demos, and they build psychological safety so people can give and receive feedback without feeling like they're walking on eggshells.

Critically, they make space for recovery. Not just after intense periods, but as a regular practice: retrospectives that lead to action, time to pay down tech debt, and permission to say "I'm at capacity" without fear of being seen as uncommitted.

When new people join and it takes them two weeks just to figure out how to run the app locally because nobody has time to help, that's a team health problem. When burnout is normalized, that's a team health problem. When only the loudest voices are heard, that's a team health problem. You can't ship your way out of these issues.

They connect effort to outcomes (and know when to stop)

Progress isn't just about movement, but about impact. Strong teams tie their work to measurable business outcomes. They prioritize based on what matters, not just what's urgent or interesting. They reflect regularly on whether their work is actually achieving results. They celebrate progress that benefits users and the business, rather than simply noting that "we shipped five things." This is the capability that separates the teams that stay busy from the teams that drive results.

A team that says "We shipped five features this sprint," might be working hard, but a team that says "We launched onboarding improvements that reduced drop-off by 12%, putting us ahead of our Q3 engagement goal," is working with purpose. It's the difference between outputs and outcomes, between doing things and doing things that matter.

All of these capabilities are essential for long-term success. But you don't need to be excellent at all of them on day one. Start by assessing where your team is strong and where they're struggling. Pick one or two areas to focus on, and build from there.

High-capability teams perform better, with less conflict, more trust, and greater impact.

Growing Without Losing the Plot

Once the capacity of a well-run team becomes the limiting factor, you've earned the right to grow. This isn't because things are chaotic or someone's asking for more hands. It's because your team is focused, capable, and delivering—you've hit the edge of what's possible without new energy or skill. That's when it's time to expand the crew.

But growth isn't linear. Adding people doesn't automatically increase velocity. A new teammate should bring more than just additional output; they should broaden the team's strengths and open space for new challenges and possibilities. The best additions expand what the team can take on, while maintaining speed.

A capable team should develop its members as it grows. This is also the moment to think about who gets to grow with you. If your team is healthy, stable, and skilled, it can become a platform for developing others. That means making space for people earlier in their careers who bring curiosity, fresh perspectives, and a hunger to learn.

Growth works in both directions. As junior or mid-level engineers gain confidence, senior team members also develop—through mentoring, modeling, and reinforcing what they know. Growth becomes mutual. The team deepens as it expands.

From Supporter to Supported

As the team grows, your job changes. You move from supporting a group to being supported by one. You stop managing every detail and start reinforcing direction. You focus less on execution and more on clarity, cadence, and culture. Before, you led by proximity. Now, you lead through leverage. This achieved through:

- Systems that support the work
- People who carry the standard
- Trust that's been earned and extended

It's a quiet shift, but it's real. You can't lead the same way with 5 people and with 20—and you shouldn't try.

You Make Yourself Replaceable

A healthy team shouldn't need you every day. That's not a problem; it's the goal. Taking time off shouldn't feel risky. Being out of the loop shouldn't stall progress. If your absence breaks the system, you haven't built a system—you've built dependency.

Sustainable leadership means building something that works without constant oversight. That's when real growth becomes possible. You can start looking further ahead—coaching instead of fixing, guiding instead of directing. And the team? They step up.

Sometimes, a tech lead starts shaping strategy. Sometimes, a senior individual contributor starts mentoring others. Sometimes, someone simply sees a gap and fills it before you even notice.

You Scale the Culture, Too

When your team is small, culture spreads naturally. People learn how things work just by being present. But as you grow, that organic transmission breaks down. Culture isn't fixed; it evolves with every new hire, process, and challenge. The question isn't whether your culture will change—it's whether you'll guide that change or let it happen by accident.

Without intentional effort, culture fragments. New people bring different assumptions. Teams develop their own ways of working. The original intent gets diluted or lost entirely. What felt natural and obvious when you were a team of 10 becomes unclear and inconsistent at 50.

Scaling culture intentionally means:

- Team agreements that evolve deliberately
- Onboarding that explains the "why" behind how you work
- Stories that preserve decision-making principles across generations of employees
- Written norms that help new people understand and uphold what matters
- Values that stay visible in daily choices as the company transforms

Culture is never "done"; it's always evolving. You must shape what it becomes, not assume it will stay aligned on its own. There's nothing wrong with the evolution of culture, but behavior often changes faster than the words written by founders in the early days. If you don't tend to culture as you grow, you risk ending up with a company that operates in ways you never intended.

The Bottom Line

Grow when you're ready, not just when you're allowed. Hire people who bring fresh energy, not just more capacity. Create space for others to grow, not just deliver. And remember: your goal isn't to lead forever; it's to build a team that can operate and make decisions independently of you.

Conclusion: From Chaos to Cohesion

Most engineering leaders don't step into high-performing teams. They step into teams under pressure, overloaded, under-supported, and unsure of what's coming next. This chapter has served as a roadmap for transitioning from chaos to cohesion—not through grand speeches or sweeping changes, but by undertaking the quiet yet essential work of creating safety, stabilizing flow, building capability, and scaling with intention. The hardest part of this job is creating the conditions where people can succeed together. That means making the team functional before it's fast, capable before it's big, and trustworthy before it's impressive.

If you do that well, you won't just lead a better team, you'll build one that doesn't depend on you to thrive. When you're ready to go deeper, others have written far more eloquently and in more detail about managing teams and scaling leadership.

In Chapter 5, we'll delve deeper into setting direction for your team.

Setting Direction

You rarely walk into a team with too little to do. More often, you inherit the opposite: a wall of priorities, each louder than the last, all of them deemed "critical."

When Juan led Stripe's LATAM engineering team, the outside story was all momentum. Headlines talked expansion: new markets, new products, new partnerships. Inside, the truth was rougher: a growing team facing 30 competing priorities, and no clear agreement on what mattered most.

Some priorities were tactical: supporting existing customers, fixing bugs, maintaining deployments. Others were strategic monsters: exploring expansion into three new countries, navigating deep regulatory compliance in Mexico and Brazil, launching new local payment methods like OXXO and Boleto, and supporting internal infrastructure and security efforts across the company. Each request came with urgency attached and each one appeared to be high priority. So, the team said yes to everything.

On paper, it looked like progress—closed tickets, shipped code, and calendars stuffed with meetings. But underneath, the wheels were spinning. Engineers were stretched thin across regulatory compliance, customer support, expansion research, and new feature development. There was wasted energy leaking in every direction and doubts arose about whether the work even mattered.

From the outside, chaos looks like momentum. From the inside, it feels like drift. Drift doesn't just waste effort; it erodes ambition. And that erosion happens to individual contributors just as much as leaders. When senior engineers can't tell which technical decisions will matter next quarter, they stop making bold architectural choices. When junior engineers don't understand how their tickets connect to outcomes, they optimize for getting through code review instead of solving real problems.

A team also doesn't earn the right to expand just by hiring more people. It has to work well first. Adding headcount without focus can make chaotic environments even worse. In the Stripe LATAM team, the breakthrough didn't come from a new process or hiring spree; it came from clarity. During a virtual leadership off-site in the height of the 2020 pandemic, the reality check was brutal: engineering leads, product managers, operations, and legal team put all 30 priorities on the table. Each topic was debated and trade-offs were made explicit. They identified which problems were existential, which could wait, and which were just distractions in disguise. The arguments were messy, but within 60 days, the noise gave way to focused execution: six priorities total, two per team.

That shift changed everything. Engineers finally had permission to say no. Meetings got shorter and focus sharpened. Work started compounding instead of scattering. For the first time, momentum on the inside matched the story on the outside.

That's what direction is—not a perfect prediction of the future, but a path that stops your team from drifting. Good direction offers a way to help people move together, creating a shield against the drift that slowly drains energy and corrodes trust.

While you'll never have all the answers, good direction means creating enough clarity to let people move with confidence and to give your team permission to stop second-guessing and hedging—and start choosing.

This chapter is about that work: why direction matters, how drift creeps in, and how to create clarity, especially when it's absent. It focuses on the cost of indecision, the power of saying no, and the discipline of consistently communicating the vision until everyone understands it.

Chaos doesn't disappear when you add more people or write more processes. It ends when you set direction.

What Causes Drift

Drift doesn't start with bad intentions. It begins when someone decides to keep options open, when a leader hedges on a priority because they're unsure. For example, drift can happen when the product team wants to test both approaches instead of choosing one, or when the engineering team splits the difference on an architectural bet so each moment feels reasonable. These approaches run the risk of becoming toxic.

The first symptom to look out for is language. Phrases like, "Let's align on this," "We'll circle back," or "We're still figuring out the details" all seem harmless. But they're actually warnings. They mean that no one wants to own the decision, or the team does not agree. So the team moves forward, burning energy on work that might not matter.

Drift picks up speed once trade-offs go "underground." For example, the sales team might promise a feature to close a deal, an exec might drop a "quick idea" in Slack, or the legal team might escalate something compliance related. Each request makes sense in isolation. But no one asks what stops to make room for the new. Priorities pile up because saying no feels riskier than saying yes. Engineers sense it first. They stop asking, "Why does this matter?" and start asking, "Which fires do I need to put out today?" That shift, from building toward something to just surviving the week, is drift taking hold. By the time leadership notices, the damage is done.

The Drift Diagnostic

When teams drift, the symptoms are often visible long before the underlying problems become critical. The diagnostic tool shown in Table 5-1 helps leaders to identify drift patterns early, and it provides concrete interventions to restore focus and momentum. Use this when you sense your team is spinning but can't pinpoint exactly why.

Table 5-1. The drift diagnostic: identifying and addressing team momentum loss

Signs of drift	Root cause	Immediate action	Sustained practice
People say, "We'll circle back," but never do.	Too many competing priorities, no clear focus	Choose one main goal for the next six weeks.	Track all work visibly. Untracked work gets deprioritized.
Teams constantly ask, "What should we work on first?"	Leadership avoiding making tough priority calls	Make a clear priority decision and explain your reasoning.	Communicate priorities repeatedly until the message sticks.
Work gets completed but doesn't build toward anything meaningful.	Team reacting to whatever seems urgent instead of following a plan	Create an explicit "stop doing" list to protect focus.	Review every meeting's purpose before it happens.
The best team members start looking for other opportunities.	Important decisions getting stuck because everyone needs to agree	Share regular updates on what you're building and why.	Create safe ways for people to disagree and ask hard questions.
The same arguments resurface in every planning session.	Teams making decisions without acknowledging what they're giving up	Make trade-offs explicit by naming what you're not doing.	Follow up after decisions to ensure commitment remains strong.

If you recognize two or more signs from the leftmost column, your team has lost focus and momentum.

Choose one action from the "Immediate action" column and implement it this week—small interventions can break drift patterns before they become entrenched problems.

The Silent Cost of Drift

Drift rarely announces itself. It doesn't look like failure. Instead, it looks busy. Tickets are closing, demos are scheduled, roadmaps are shuffled forward, And standups are full of updates. On the surface, things seem to be happening. But underneath, urgency has replaced clarity. People are sprinting, and no one can explain why.

That's drift: when effort no longer compounds. You hear it first in a standup meeting. Someone says, "Let's just get something out" without explaining why it matters. Another engineer mentions they'll "clean it up later," but you both know later never comes. A product manager says, "The product team is still figuring it out," and no one pushes back because everyone's tired of asking. These can be indicators of what people say once they've stopped believing that the work they do is tied to anything real.

At first, drift feels like background noise. But drift compounds. Every unclear trade-off doubles the work, every abandoned priority leaves damage, and every shifting goal erodes trust. By the time morale collapses, the damage is already baked in.

And the cost is brutal. Execution slows because no one knows the absolute priority. Trust erodes because decisions don't stick. Engineers stop taking initiative because they can't tell what matters. Product managers optimize for marketing instead of impact. Leaders run meetings full of "updates" that don't add up to progress. Top performers—the ones you count on to steady the ship—burn out fastest.

Juan saw this in the Stripe LATAM team before the off-site leadership ever took notice. The team was stretched across expansion research for three new countries while trying to deliver compliance requirements in Mexico and Brazil. Engineers spent days on regulatory documentation that might not matter if expansion priorities shifted. Others built payment method integrations that stalled when partnership negotiations changed. Global customers needed regional support, but the team couldn't predict which requests would become long-term commitments. Energy scattered across too many fronts, none getting the focus needed to finish well. It only takes a few cycles like that before even the most motivated people shut down.

Engineers stopped pushing bold ideas and instead optimized for survival: "Why bother cleaning this up if they'll change direction next week?" Designers stopped asking for the reasoning behind decisions and just shipped what would get approved.

Managers stopped coaching and started counting tickets, because tickets were the only thing that felt measurable. People didn't quit all at once, but they quit piece by piece, until the culture hollowed out.

In Practice: April's Story

At PixelCurl, April inherited the aftermath. Two senior engineers had left in the previous quarter. Their exits weren't dramatic, but they just quietly accepted offers elsewhere. The junior engineers who remained had stopped asking questions in standups because "Nobody knows what we're supposed to be doing anyway." One designer told April in a one-on-one, "I used to propose ideas, but now I just wait to be told what to build."

The real damage isn't wasted time; it's what happens to the people. From the outside, dashboards stay green, demos look polished, and executive updates are full of slides showing "progress." Leaders confuse activity with momentum, so by the time they notice the problem, the team is already exhausted.

Teams don't collapse from one catastrophic failure; they erode, developing feelings of burnout, cynicism, and misalignment. People don't leave because of one bad sprint—they leave because months of unacknowledged and unresolved drift convinces them that their work doesn't matter.

But here's what gives us hope: the opposite is just as true. Even a little clarity can flip a team's energy. Drift drains momentum invisibly, but direction restores it instantly.

What Good Direction Looks Like

Drift shows you what happens when direction is missing: teams burn energy without progress, motivation erodes, and even the strongest people check out. But the inverse is just as powerful. When direction is clear, the same team that looked exhausted can rediscover momentum.

If drift is motion without meaning, then direction is focus with momentum. It doesn't make the work easy but it makes it make sense.

Stripe's LATAM payments team had been stuck in the same cycle—too many priorities, endless context switching, and constant noise. Then leadership made the call: narrow their efforts to shipping OXXO in Mexico, and Boleto in Brazil. Once the LATAM team forced priorities into the open, the team's posture changed almost overnight.

Senior engineers could finally challenge ideas without sounding cynical. Before, every debate felt endless. After all, why sharpen work that might be dropped tomorrow? Afterward, pushback was welcomed because everyone knew which trade-offs mattered.

Junior engineers felt safer because they weren't coding tickets into a void anymore. They understood how their work connected to outcomes in Mexico and Brazil. That clarity gave them confidence, and confidence accelerated learning.

Product managers stopped competing over priorities. Previously, priorities shifted so quickly that project managers jockeyed for attention, trying to get their features approved before priorities changed again. After, they were aligned with engineers on the same outcome. They could defend choices, say "no" without damaging credibility, and trust that decisions would stick.

Team leads felt relieved. They weren't the bottleneck anymore; they no longer needed to referee every trade-off or escalate every disagreement. With context, the team members made smart calls themselves.

Direction changes not just execution, but how people act. You can hear it in the way they talk. Someone who used to say "We'll just get something out" starts saying "If we add this scope, we'll miss the Brazil deadline." Status updates that used to be vague and apologetic are now tied to outcomes, such as "This release unlocks the top payment methods in Mexico." The shift isn't cosmetic; it's people finally understanding what they're building toward.

Meetings changed, too. Before, meetings consisted of endless debates, circling without resolution. After, meetings held sharper discussions, where trade-offs were named out loud and decisions were made that actually stuck. Slack threads got shorter, context switching dropped, and the energy that was once spent negotiating priorities went back into building.

The technical challenge was still high—compliance projects are always tricky—but morale flipped. People leaned in again, taking pride in progress. The same team, the same talent, was suddenly compounding instead of dissolving. That's what good direction looks like—not perfection or certainty, but just enough clarity for people to move with confidence.

Clarity can feel like a constraint in the moment. But in reality, it's freedom. Saying no to everything that doesn't align gives people permission to move fast and finish. The Boleto scope decision turned endless ambiguity into a clear priority: basic payments first, enhancements later. By making the limits explicit, the team finally aligned on what would happen, rather than focusing on what wouldn't.

Good direction doesn't remove ambiguity or trade-offs, but it makes ambiguity manageable and trade-offs clearer. You don't need everyone to agree with the decision; you just need them to commit once it's made.

When direction is sharp enough, you see a consistent transformation: senior team members make clearer decisions instead of revisiting them, junior team members grow faster because they feel safe to ask and learn, product managers defend priorities instead of competing for attention, leaders step back from refereeing every debate and, most importantly, teams compound progress instead of dispersing effort.

Direction is what unlocks autonomy and makes teams stronger. But most organizations don't get there on the first try. They stumble into the same traps again and again. If you learn to spot them early, you can correct course before drift takes over.

Traps That Cause Drift

Most teams don't fail because they lack talent. They fail because they fall into predictable traps. Oftentimes, chaos merely magnifies the problems that were already existing.

No Clear Inputs

Sometimes there isn't a real strategy at all. Or worse, there's a 40-slide deck that says everything but commits to nothing. Teams sit in limbo, waiting for clarity that never comes.

At Stripe LATAM, this showed up as competing signals from different stakeholders. Global customers needed regional payment support. Expansion research suggested opportunities in three new markets. Regulatory compliance demanded immediate attention in Mexico and Brazil. Each signal made sense in isolation, but together they caused major sticking points. Engineers guessed at priorities. Managers stalled, hoping leadership would clarify. Nobody wanted to make architectural decisions because the strategy kept shifting. For the team, it felt like running in place. For leadership, it felt like "keeping options open." It was drift in disguise.

To avoid this, don't wait for perfect clarity from above. Set a local mission, even a temporary one: "Here's what matters for the next six weeks." That gives your team a compass, even if the bigger map is still being drawn.

Too Many Inputs

The opposite problem is just as cumbersome. For example, within the context of our LATAM scenario, we could have multiple stakeholders with competing priorities, or global customers promising big deals if regional payment methods ship fast. In addition, legal teams might be escalating compliance issues that could shut down markets, partnership teams could be pushing integration deadlines, and internal infrastructure teams might need LATAM support for security audits. The backlog becomes a landfill of competing urgencies.

Stripe LATAM lived here early on—with no less than 30 priorities, all labeled "critical." Engineers felt whiplash between regulatory work and feature development. Product managers played politics, trying to get their initiatives prioritized. Everyone tried to get their request to the top. The fix wasn't a process improvement like a new project management software. Resolution came when making trade-offs explicit during a virtual leadership off-site. Leaders wrote down everything the team had been asked to do, and then chose. They created dependency maps, "not doing" lists—whatever it took to show that every yes had a cost. That's how 30 priorities became 6.

Hesitant Leadership

Sometimes leaders see the chaos but won't step in. They confuse flexibility with indecision, and so they wait for certainty. Meanwhile, the team drifts.

Before the LATAM off-site, Juan watched leaders quietly acknowledge the chaos but hesitate to make hard calls about market expansion versus regulatory compliance. They chose to delay rather than risk blocking growth opportunities. But inaction is its own decision—and it's usually worse. For engineers, it felt like abandonment. For leaders, it felt like keeping options open. But it was neither. Clarity doesn't require certainty. Sometimes, you just have to make the call, explain the why, and own the outcome. Even if you adjust later, you've given your team clear direction.

Reactive Roadmaps and Consensus Traps

Two traps often travel together. The first trap is roadmaps that shift with every escalation. Teams stop building toward outcomes and just react to the loudest voice. Sales demands a partnership feature. Marketing wants community features. Leadership wants payments compliance fixed. Before the LATAM off-site, directors waited for executives to force choices instead of committing themselves. Different stakeholders pushed competing timelines without owning the trade-offs.

The second trap is the opposite: the endless search for consensus. Leaders wait until everyone agrees before deciding. This results in meetings that drag and seemingly endless debates. Nothing moves.

Stripe avoided both when it chose to focus on Mexico and Brazil. The decision wasn't unanimous. Some leaders argued to keep expanding. But once the call was made, everyone committed, and that is what mattered—not agreement, but commitment. Debate until the decision, and then support it.

The way out of reactive chaos isn't more meetings; it's anchoring to outcomes. Instead of chasing features, define goals (e.g., "Improve reliability by 50%" or "Onboard 1,000 merchants"). Then, measure every new request against those outcomes. Without an anchor, every wave knocks you off course.

These traps are predictable once you know them. Spot them early, and you can course correct before drift takes over. But not every threat to direction shows up this obviously. Sometimes the danger is quieter: the team looks productive, roadmaps shuffle forward, updates sound positive, but the work does not advance the real goals. That's when leaders need to be the lighthouse.

Be the Lighthouse

After the obvious traps—too many inputs, hesitant leadership, and reactive roadmaps—the hardest part is leading teams once drift is already happening.

Once drift sets in, what teams need most is a fixed point: a clear reference they can align to when they've lost direction. That's your role as a leader: to be the lighthouse they can steer toward.

Drift doesn't announce itself with alarm bells. Calendars stay full. Tickets move. Updates sound positive. From the outside, it may look like momentum. But from the inside, people are guessing. An engineer suggests we "refine it next sprint" because they're not sure what completion looks like. A project manager says, "We'll know more once we ship," which really means "I don't know if this matters." Someone invokes "this came from the top" to end the debate instead of clarifying it. Individually, none of this is catastrophic. But together, it means goals aren't clear enough to guide decisions.

Drift appears in the backlog. Every idea, escalation, or quick win lands without context. Nothing is evaluated against a single outcome. The backlog stops being a prioritization tool and becomes a seemingly never-ending list of tasks that no one seems to get to.

Drift also reshapes behavior. Engineers stop asking if the work will matter and settle for whether it will pass review. Product managers become translators of competing requests instead of stewards of trade-offs. Leaders avoid providing clarity because they can't promise certainty. The team adapts to the drift rather than fighting it and correcting it.

The emotional toll is real. The way that drift works is that instead of sparking outrage, it slowly dulls energy. People stop proposing bold ideas because they aren't confident that the work will have an impact. They limit their ambition to what can survive review, sprint by sprint. It feels safe, but progress is slow.

The solution isn't a grand strategy deck. It's a clear outcome and a short horizon—a fixed point to navigate toward.

The Stripe LATAM breakthrough happened when leaders created clarity during that off-site and defended it through months of competing requests afterward. The team knew why their compliance work mattered, even when it was tedious. They understood how OXXO and Boleto connected to regional growth, even when the technical challenges were brutal.

The Boleto payment method also circled for months under a pile of reasonable requests: advanced fraud detection, multiple bank integrations, enhanced user verification, and adoption analytics. Every feature made sense for the Brazilian market. Together, they delayed the launch indefinitely. The drift broke when the team decided to focus on core functionality first: basic Boleto payments that worked reliably. Enhanced features would come after adoption was proven. So, the scope shrank, energy rose, and Boleto launched. Enhanced features were shipped in subsequent releases based on actual usage data. Brazilian customers got value faster, and the team rediscovered momentum.

In both the Stripe LATAM prioritization and the Boleto scope decision, clarity came from framing near-term outcomes and repeating them until they felt obvious. It wasn't a perfect map, but a clear point on the horizon.

In Practice: April's Story

April did this at PixelCurl by picking one fixed goal: ship a stable rendering pipeline for their top five clients by the end of the quarter. Everything else—the mobile app, the infrastructure rewrite, the experimental features—waited. She repeated it in every standup, every planning meeting, and every hallway conversation. Engineers initially pushed back, asking about the other priorities, but she kept redirecting: "Does this help us ship the rendering pipeline? If not, it's not in scope this quarter." By week three, the team stopped asking. By week six, they were moving faster than they had in months.

If you inherit a drifting team, start small:

- Name one outcome for the next cycle in plain language. Avoid vanity metrics; pick a goal that guides your decisions.

- Write a not-doing list and make it visible. Update it whenever you say no.

- Anchor status updates in one sentence: "We are doing *X* so that we can achieve *Y*."

- Create deadlines around uncertainty. If you can't decide now, then set a date and define what will be true by then.

Drift doesn't correct itself. You don't wait for clarity to emerge. You create it by setting clear goals and repeating them consistently. Be the lighthouse.

Creating Alignment

Setting direction inside a team is one thing. Making it stick across teams, functions, and leadership levels is another. Alignment is where most strategies fall apart—not because people are lazy, but because smart people optimize locally without understanding the broader context.

Why Alignment Breaks

Misalignment rarely looks like sabotage. It often appears as if everyone is doing their job. Product managers chase customer satisfaction scores. Engineers chase technical elegance. Sales push revenue targets. Finance pushes margins. Each choice makes sense within that function's frame of reference. But without a shared context, those choices cancel each other out.

Before the Stripe LATAM off-site, misalignment showed up more subtly. Different stakeholders optimized for their own metrics without seeing the bigger picture. For example, the partnerships team pushed integration timelines that conflicted with compliance requirements, and customer success promised features to close deals without checking engineering capacity.

The cost of these individual decisions was measurable: new payment method launches were delayed by months, as teams worked on conflicting assumptions. Engineers couldn't tell which regulatory requirements were firm deadlines versus aspirational goals. Customer features were half built because priorities shifted before anyone could finish the compliance work required to launch them.

Alignment Is Not Agreement

Alignment doesn't mean consensus. It doesn't mean everyone loves the decision. It means disagree and commit anyway: the decision stands, and everyone must commit their team's full effort to making it work.

At Stripe, the LATAM team faced a brutal trade-off: stabilize Mexico and Brazil first, or push expansion into three new countries. Everyone wanted growth. No one wanted to pause. Expansion was exciting, but stabilization felt like grinding compliance work. The debates were heated. But once the trade-off was named and the call was made, the team committed. Stabilize first, and then expand later. That choice gave the team clarity and the business real leverage in those markets.

Conflict in the open is healthy. Silence after the decision is dangerous. You want engineers to be able to say, "This timeline ignores regulatory testing; we will fail if we don't adjust." You want product managers to feel comfortable saying, "This request derails our focus and delays the Brazil launch by two weeks." Expressing disagreement clearly will improve the work and surface real constraints. Pretending everyone agrees will slow progress and create problems later.

How Leaders Create Alignment

Alignment doesn't happen by accident. Leaders make it happen through deliberate practices. Some of these practices are outlined here:

Translate the mission into something concrete.
> "Expand across LATAM" is too abstract. "Enable the first 1,000 merchants in Brazil to accept payments" gives everyone a clear, measurable goal. Junior engineers understand how their work connects. Senior engineers know which technical decisions matter. The goal becomes tangible.

Make the trade-offs visible.
> List everything openly. Use dependency maps, not-doing lists, or any tool that exposes hard choices. When the LATAM team listed all 30 priorities, the cost of saying yes to everything became impossible to ignore. The choice between breadth and depth was evident.

Close the loop after the debate.
> Encourage vigorous discussion until the decision point, and then conclude it: we have argued, weighed the options, and chosen—now we execute as one team. Run a commitment check. Ask each leader to state what the decision means for their team's work. If the answers diverge, you're not aligned yet.

Shut down hallway politics immediately.
> The fastest way to slow progress is letting side conversations rewrite decisions made in the meeting. When someone starts undermining a call in private, address it directly: "We made this call together. If you have new information that changes it, let's discuss it with the whole team. Otherwise, I need your full commitment."

Show the impact of misalignment.
> Help people see the effect on customers when features are delivered incomplete. Show the cost to the team when senior people quit because they can't tell what matters. Make local optimization costly by revealing its impact across the organization.

Alignment isn't overhead. It's the foundation that lets teams move fast without falling apart. It ensures that conflict improves decisions instead of dividing the team, that trade-offs stick instead of drifting, and that progress compounds instead of scattering.

Adapting Without Losing the Plot

Even with clear direction, plans will change. Competitors move. Regulations shift overnight. What mattered last quarter may not matter now. Change itself isn't the problem; the problem is unexplained change.

Unexplained change hits everyone differently, but it always erodes trust. Engineers feel whiplash. Product managers feel like they just broke a promise to their team. And leaders? They often think silence is safer than admitting uncertainty. It never is. Unexplained change erodes trust faster than bad news delivered clearly.

At Stripe LATAM, the pressure to expand was intense. Leaders wanted to open three new countries immediately. On paper, it looked like unstoppable momentum. On the ground, the team was barely stable in Mexico and Brazil, and regulatory obligations in both markets were piling up faster than they could address them. Expanding further would have created a compliance crisis that could shut them down entirely.

The hard call was to focus on stability in Mexico and Brazil before expanding elsewhere. The message was direct: the mission remained regional growth, but the path was compliance and adoption first, then expansion. New markets would wait until the foundation was solid.

Not everyone agreed. Launches were exciting and generated good press. Stabilization involved tedious compliance work that no one would notice until a problem arose. Some leaders pushed to keep expansion on the agenda, arguing that competitors would gain ground. Others quietly admitted relief; they'd been worried about the technical debt and regulatory gaps but hadn't wanted to be the one to slow things down.

But once the trade-offs were clearly stated and the business risk explained, the decision held. The plan changed, but the core mission stayed consistent: build the foundation that enables the team to succeed in the region.

If the Plan Changes, the Story Stays Intact

Plans change naturally as you learn, but stories endure as your North Star. If people understand why the plan shifted and how it connects back to the mission, they can accept it and adapt their work accordingly. If they don't understand the connection, the same shift looks like chaos or failure.

Direction isn't about eliminating change. It's about creating a strong enough framework for the team to adjust without losing coherence. When the story—the mission—is clear and consistently communicated, people can handle plan adjustments. But if the story keeps changing, everything falls apart.

Adaptation proves that clarity isn't a one-time achievement. It has to be renewed every time plans bend, priorities shift, or new information emerges. Clear direction ensures the team can continue working effectively through uncertainty.

And that points to the deeper truth running through this entire chapter: clarity isn't extra work; clarity *is* the work.

Clarity Is the Work

Teams don't burn out from hard work. They burn out from unclear work. Long nights and strict deadlines are manageable if people understand why the effort matters. What drains them is pouring energy into tasks that shift, change, or disappear without explanation.

Direction is the scaffolding of leadership. Without it, even the most talented teams drift—busy, exhausted, and unsure if their effort matters. With it, those same teams compound progress and build momentum that survives setbacks.

At Stripe LATAM, clarity meant putting 30 competing priorities on the table during that virtual leadership off-site, naming trade-offs out loud, and forcing leaders to choose. That single process unlocked autonomy and gave engineers permission to say no to distractions.

For Boleto, clarity meant setting a scope boundary: core payment functionality first, and add enhanced features only after adoption was proven successful. That constraint turned a stuck project into a shipped product. Features that had been debated endlessly suddenly had an answer—if it's needed for basic payments, we build it; if it's an enhancement, it waits for future versions. The clear scope enabled the team to execute effectively.

Clarity also lives in team habits:

No ticket, no work
>Tasks not in the tracker aren't considered active work. That single rule filters out noise before it reaches engineers.

The purpose of this meeting
>Start every call by stating why it exists. If there is no purpose, then there should be no meeting.

Asking anything
>Keep an open channel where questions can be asked without fear and where doubts can be addressed.

None of these practices is glamorous. They won't get you headlines or praise from executives. They may feel tedious. That's the point. Clarity isn't a heroic moment; it's the discipline of repetition.

Leaders often tire of saying the same thing. They assume the message has landed because they've said it once, but it rarely has. The truth is that when you are sick of repeating yourself, your team is only beginning to believe you. Repetition isn't laziness; it's leadership.

And clarity isn't necessarily about certainty. You will never have perfect information or be able to see the whole map. Clarity is about giving people a sharp point to navigate toward, even in uncertainty.

In addition, complexity doesn't disappear and conflict doesn't stop. Uncertainty is still there, but now people can navigate through it instead of drowning in it. Conflict sharpens decisions instead of fracturing teams, and uncertainty stops being an excuse to delay. You need a clear point that people can aim for, rather than a perfect plan.

In chaos, clarity is the rarest and most valuable gift you can give. It turns drift into direction, exhaustion into energy, and noise into momentum. It won't feel glamorous or revolutionary. But it is the work that makes every other kind of work possible.

So, we repeat: clarity isn't extra work; *clarity is the work.*

Exercise: The Clarity Workshop

Block 75–90 minutes with your team. Make sure all phones are down and laptops are closed. Depending on your setup, use one wall, board, or shared doc. The goal is establishing a sense of alignment. Use the following structure to guide your workshop:

1. *Map reality (20 min):* Start by naming what's true right now: current projects, deadlines, urgent issues, and unspoken frustrations. Write it all down without spin. Focus on honesty, not presentation.

2. *Choose the sharp point (15 min):* Identify one or two outcomes that matter most this cycle. How will you know if you succeed? Encourage debate until the outcomes are clear. If the team can't agree, that's valuable information; surface the disagreement and resolve it before moving forward.

3. *Draw the not-doing line (15 min):* Make a visible list of the tasks that you are explicitly cutting or deferring. Every yes has a cost. This step is where you protect the team's energy.

4. *Craft the story (10 min):* Write one sentence, simple enough for anyone to repeat: "We are doing X so we can achieve Y." Say it out loud until everyone understands it clearly.

5. *Commit together (10–20 min):* Go around the room. Each person states what the decision means for their work. If answers diverge, stop and clarify until everyone is aligned.

The artifact doesn't matter. What matters is the conversation and shared commitment, not the notes on the board.

Reflection

Take a moment to consider the following reflection points:

- Which piece of work is your team doing right now that is drift (motion without purpose)?

- What is one outcome you could name that would remove that drift?

- If you had to defend your not-doing list in front of an executive, would you have the conviction to maintain it?

Figure 5-1 illustrates the drift recovery cycle.

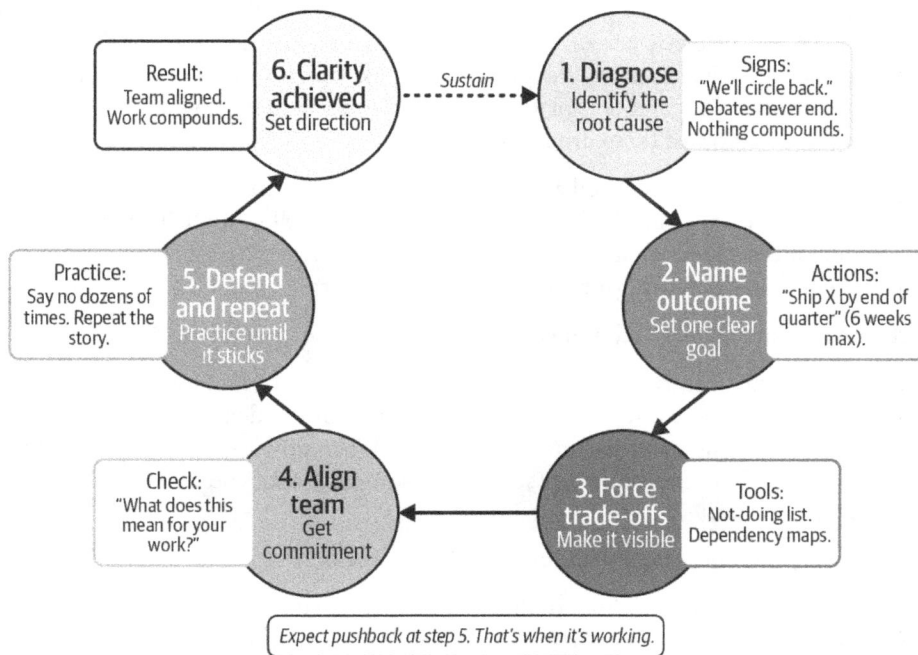

Figure 5-1. *The drift recovery cycle: from detection to sustained clarity*

Conclusion: Pulling It All Together

The hardest part of setting direction isn't the decision itself. It's maintaining it long enough for it to guide the team.

At Stripe LATAM, after the team finally cut down priorities, the pressure to add more never stopped. Sales requested exceptions, executives pitched quick wins, and partners escalated urgent requests. Every week brought a new reason to say yes, to keep options open, or to hedge just a little.

What saved the team wasn't a better process or a stronger roadmap. It was the leaders who kept saying no—not once, but dozens of times, in standups, in planning meetings, in hallway conversations. The repetition felt boring but it was the only thing that worked.

Most leadership advice treats direction like a light switch: flip it once and the room stays bright. But real teams are messier. Direction is something you set, then reset, then defend, and then explain again. If you think you've said it enough, you haven't. When you're tired of repeating yourself, your team is just starting to believe you.

The other thing no one tells you is that setting direction makes you a target. When you say no to one priority, you're saying no to someone's project, someone's promise, someone's bet on what matters. People don't thank you for clarity. They may push back, work around you, and complain that you're slowing them down. Some of them may be right, but most of them aren't. Your job is to hold the line anyway.

Because here's what we've learned after years of observing teams drift and recover: the teams that survive chaos aren't the smartest or the most well funded. They're the ones where someone decided what mattered and refused to let it slip. They set a direction, made it visible, and defended it past the point of comfort.

Drift is the default state. Direction is the work you choose to do instead.

If your team is drifting right now, start smaller than you think. Don't use a strategy off-site, or a new OKR framework. Pick one outcome for the next six weeks and make it so clear that anyone on the team can repeat it. Write down what you're not doing and put it somewhere everyone can see. When someone asks you to add more, point to the list and say no.

That's it: one outcome, one not-doing list, and six weeks.

You don't need permission to do this. You don't need executive alignment or a perfect plan. You need the conviction to create clarity when no one else will, and the stubbornness to defend it when people push back. The rest follows from there.

In Chapter 6, we'll talk about how to prioritize your new direction.

Shipping Products and Code in Chaotic Environments

We discussed in Chapter 1 that doing something is (almost) always better than doing nothing. We stand behind this take but acknowledge that it's not very nuanced, because there is often more that you should or would do if you had the resources to. So, this means that you must decide *what* you will do and *the order* in which you will do it.

To do this, you need to consider prioritization approaches and exactly how to determine your actions. However, circumstances in most chaotic situations don't lend themselves to implementing highly structured processes and techniques. Indeed, applying those more elaborate processes in a chaotic environment often means serving the process instead of driving the outcome. Don't underestimate the power of simplicity and common sense.

We need to find some techniques that meet the following criteria:

- Inspire confidence.
- Be execution-focused.
- Be lightweight and adaptable.
- Foster continuous learning.
- Focus on communication and collaboration.

In this chapter, we explore what this means and introduce you to some approaches that could either meet your needs or be adapted to meet them.

This won't be a detailed guide to prioritization and the software development lifecycle (SDLC). But it does offer a crash course designed to equip you with the skills to manage right now. For deeper guides, check out Martin Fowler's collection of Agile guides (*https://oreil.ly/-t_5K*) and Atlassian's SDLC guide (*https://oreil.ly/ACdjL*).

Inspire Confidence

You need to make your team confident that processes will work most of the time; here, we're fond of the 80/20 rule: if a process works 80% of the time, it's a success. Additionally, every successful process iteration will give the team more confidence and reassurance that they are improving. Growing confidence, or at least not eroding it, is also achieved through admitting that processes can be failures—and making those failures learning opportunities.

Here, you can also start small and build up. For example, in many chaotic environments, estimation is a scary prospect. The question of how long something will take—especially if those estimates are being communicated to leadership or other teams—often results in much debate and churn. You will inevitably get those initial estimates wrong. When that happens, the team will likely lose confidence in their ability to deliver and in your ability to lead.

Instead of only making a wild guess, acknowledge uncertainty and then put a stake in the ground. For example, you could adjust your initial tolerance to accept a 50% variance in an estimate: if your team thinks it's a week of effort, it could be two weeks. It's still a guess, but you acknowledge that you don't know what you don't know. Admitting uncertainty and articulating an approach to add more rigor and help reduce that uncertainty is much more confidence-inspiring than throwing arbitrary estimates out. This also drives a desire to learn, measure, and refine the process, hopefully reducing that variation over time as you gain experience.

In chaotic environments, you want to instill confidence and reduce fear. The best way to do this is to have quick, iterative wins that grow your and the team's confidence. In a prior role, James inherited an engineering organization stuck in a cycle of constantly changing priorities, incidents, and temporary fixes. To counteract this situation, the team had stacked processes onto their lifecycle to gain some control of the problem. As a result, the whole software development process had ground to a halt. James stripped away the process to address this and started looking at what mattered to this team at their current level of maturity: shipping quality code. After some initial shock, the team began to ship code. The mere experience of moving forward and delivering gave the team enormous levels of confidence. The code was far from perfect, but the forward progress—after months of stagnation—changed the whole mental outlook of the team.

We can return to April at PixelCurl as an example of how this might pan out.

In Practice: April's Story

Similarly, April at PixelCurl faced a confidence crisis with the StudioOps team. Senior engineers working in silos had created an environment where junior engineers felt unsupported and isolated. April's first priority wasn't to add complex processes but to rebuild confidence through small wins. She started by pairing senior and junior engineers on a quick bug-fix sprint—nothing revolutionary, just getting people to ship code together. The shared success of clearing a backlog that had been lingering for months gave both groups a confidence boost and helped rebuild collaboration between senior and junior engineers.

Importantly, you must also acknowledge your losses and turn them into learning and improvement opportunities. Confidence comes from knowing there is direction—that someone has a plan and is at least trying to drive improvements. Grandiose promises and complex frameworks rarely achieve the same result.

Be Execution-Focused

We must also ensure our chosen processes focus on execution and outcomes. Any process that prioritizes the outcome in favor of the process is a red flag—for example, when conducting a ceremony becomes more important than actually shipping the product, or worse, when teams spend time arguing over or endlessly fine-tuning processes. There's a specific type of person—usually the one who reads the rules intensely when playing a board game—who lives to create processes and, more dangerously, to identify every possible edge case and insist the process must address them. Other people cling to process as an anchor amid the chaos around them or as a way to mitigate risk.

As leaders, we and our teams are all ultimately measured by deliverables, and we must keep that in sight in every task we undertake. "We built a comprehensive process that addressed every edge case" is meaningless unless the process results in a successful deliverable. The same goes for "the process worked"—if it didn't achieve the deliverable, it didn't work. We like to approach process design from the perspective of the deliverable and work backward from there. We know what we want as a deliverable—for example, shipping a release, a product iteration, or delivering a feature—and then layer in the bare minimum process necessary to help the team achieve it. In "Prioritization Techniques" on page 101, we explain how to build those layers, and we provide examples and models designed to be execution-focused.

Be Lightweight and Adaptable

This leads us to the importance of choosing lightweight processes. One of our favorite maxims is: *You can always add a process, but it's much harder to remove it.*[1]

So, we always start with the bare minimum of process necessary to execute. We focus on guardrails to ensure the team is focused on what needs to be done and executing on that. Your mission in building the process is to provide those guardrails and the resources necessary to execute, and to help remove any friction or obstacles in the way of that execution. This mission is much easier if you let the process grow from a seedling rather than attempting to plant a forest.

Kevin Stewart, an engineering leader at several startups, paraphrases Michael Pollan's *Food Rules* to describe what that process should look like (*https://oreil.ly/g6cQO*).

Michael Pollan's food rules:

> Eat food.
> Not too much.
> Mostly plants.

Become:

> Use a process.
> Not too much.
> Mostly agile.

Let's break down these rules. First, use a process: you must have something. Many early-stage companies are collections of individual contributors working separately on a product. If you're working with a few smart engineers building a minimum viable product (MVP) with a narrow focus or an obvious set of objectives or features, this approach often works—sometimes even moderately well. It also works when two to four folks sit around a table in a coworking space. However, it doesn't work if you need to make decisions at a larger scale about what is a priority. And it doesn't work as you scale the size of a team or company or if you need to collaborate with other people or teams. Your role is to build up your team's capacity, capability, and confidence. To do that, you'll need to establish a few simple structures.

Secondly: not too much. We've discussed why this is important. Too much process leads people to serve the process rather than deliver the outcome.

---

1 Why is removing processes hard? People fear losing perceived safety nets. Dependencies form around processes, and no one wants to be responsible for what might break when they're gone.

Thirdly: mostly agile. You'll note the "mostly," and that "agile" here is not capitalized. That's a deliberate decision by Stewart. It's important to highlight that the term agile (or Agile) now encompasses an extensive variety of practices and disciplines. Some are simple, lightweight processes, and others are multipage volumes that make the assembly instructions for a Lego Death Star™ look simple. However, we still want to choose some principles, practices, and processes from the Agile ecosystem.

There is a more metaphysical question that many ask at this point: why Agile at all? You could argue a lot of perspectives here; there's extensive data and analysis of the success and failure of various Agile processes, but we like to summarize it as "the least bad approach to building products." (We discuss other frameworks for problems outside of product development later in the chapter.) That might appear oversimplified, so let's fill in some blanks. First, the original Agile Manifesto is over 20 years old; there has been lots of movement since then. Second, consider the work of the original manifesto authors (*https://agilemanifesto.org*):

> We are uncovering better ways of developing software by doing it and helping others do it. Through this work, we have come to value:
>
> - Individuals and interactions over processes and tools
> - Working software over comprehensive documentation
> - Customer collaboration over contract negotiation
> - Responding to change over following a plan
>
> That is, while there is value in the items on the right, we value the items on the left more.

Some of that might sound familiar to anyone who has gotten this far into the book. We believe in many of the same values articulated in the manifesto. The final line is especially resonant because its interpretation has driven much of the change in and adaptation of Agile over the years. The relative value placed on the right and left items often shapes the specific Agile approaches to building software. But we've also built software under less-than-ideal circumstances in chaotic environments, so we apply some pragmatism to the manifesto's values and the weights applied to each side.

That's where "mostly agile" comes into play. Rather than dogmatically choosing to "do Agile," selectively adopt the practices that work for you and your team, apply them, and put aside the rest; you may find some useful later. The real power of agile isn't in following every ceremony to the letter, it's in embracing its core principles of people, shipping software, collaboration, flexibility, and responsiveness. This is especially true in chaotic environments where you need to inspire confidence in your team, adapt to rapidly changing circumstances, and make decisions without perfect information.

We're not going to explain Agile development in any significant depth, as there are a lot of resources out there that can do that for you. Some good starting points are:

- Martin Fowler's collection of Agile guides (*https://martinfowler.com/agile.html*)
- *The Art of Agile Development* by James Shore and Shane Warden
- *Essential Scrum: A Practical Guide to the Most Popular Agile Process* by Kenneth S. Rubin
- *User Stories Applied: For Agile Software Development* by Mike Cohn

We will, however, express our opinions about what bits of Agile work and how to configure them in chaotic environments. We continue to emphasize the importance of experimenting and finding a process model and cadence that works for you and your team.

Choosing the right pieces and level of process also helps you drive the iterative development of how your team ships software. Lightweight, simple, and adaptable processes lend themselves to experimentation and continuous improvement. If you try something new as an experiment, but it doesn't work, then you can iterate to something new and try that until you find the process that fits, or so that you can scale your process as your team grows in confidence, capacity, and capability.

Foster Continuous Learning

Any new process is a learning opportunity. When you implement a new process, think about a way to measure its efficiency and efficacy. Implementing a process without the means to measure it is a wasted opportunity. Those measures can be qualitative or quantitative (or both). In the agile world, a qualitative measure could be a retrospective, the team's state of mind, while a quantitative measure could be the percentage of stories completed in a sprint against the plan. In exploring potential methods, we highlight how each can be measured and what to take from those results.

Measurement also supports experimentation. We need to be adaptable and flexible. Running measurable experiments in process provides comfort that we're making the best possible decisions, or at least decisions supported by data. So, collect feedback, measure outcomes, and refine processes over time.

Focus on Communication and Collaboration

Remember that new or altered processes are changes. And change is scary for people, especially when it's change tied to how we do our work. Best practices to ensure people feel more comfortable with change include:

- Open, transparent communication
- Collaboration
- Not doing too much at once

It's crucial to communicate changes to your team. Keep your team informed of the progress of the process and any proposed changes. Be transparent and honest with them about the challenges you are facing, what's going well, and what's not. Odds are that they already know what's working and what's not, so making any measurements, challenges, and success transparent to them grows their trust and confidence in you. Build process methodically; introducing too much at once will mean people will feel overwhelmed. So start small.

When you're building processes, remember that you are not the only person affected by them, and you're also not likely to be the one executing them every day. Go to the source and work directly with your team to find processes that work for them and you. The people running the process daily often have valuable insight into opportunities and challenges, and involving the team in developing the process will increase their sense of ownership, control, and confidence in it.

Celebrate

Remember to celebrate your successes. As much as we focus on improvement, iteration, and turning failures into opportunities, we should also acknowledge when things go well. If you have a successful week or sprint outcome—no matter how small— acknowledge it and celebrate it. Confidence in a new or changed process is fragile, especially in chaotic and unpredictable environments. So, bolster that confidence by recognizing when your team is successful.

Other Prioritization Concerns

Additionally, we have three further concerns when considering prioritization approaches:

- Not all work is created equal.
- Not all work can use the same process.
- Not all work is known at the start of prioritization.

Not All Work Is Created Equal

What do we mean by saying that not all work is created equal? Prioritization is rarely single threaded; your team will likely do many different things. Seldom can you produce a stack-ranked queue of work that your team must do. Work comes in various sizes and has different skill requirements, dependencies, and difficulty levels.

When thinking about scheduling work for my teams, we think more about working on a jigsaw puzzle than imagining it as a queue: what fits where, with what else, with whom, and in what order.[2] We need to select processes that cater to this variation without requiring enormous overhead—both people and operational overheads—to track them. Indeed, as we've discussed when we talked about the myriad roles you might need to adopt, you and your team will likely be the program and project (and sometimes product) managers for most of the work you're undertaking.

Not All Work Can Use the Same Process

Many people try to apply a uniform methodology to every task they undertake. We often see this in organizations that adopt some variant of Agile methodologies. Agile is well suited to smaller iterative tasks, but some projects don't fit this approach. For example, moving an office lends itself much more to traditional project management methodologies, as they can have complex dependencies and very linear timelines. There is often an inherent skepticism about the efficacy of waterfall-style projects—and in our opinion, rightly so when considering building software products. However, for some projects, like an office move, traditional project management can be the right choice, and you should have the ability to plan and manage a project like this effectively.

Not All Work Is Known at the Start of Prioritization

A common analogy for describing new or changed work that appears after the project starts is: building the plane while it's in flight. James is not a good flyer, so he does not enjoy this analogy. It also galls him because most of the time, none of us are building airplanes, whose design and construction—some recent examples aside—have considerably higher tolerances and standards than creating a marketing website. But the point is that the reality of many chaotic environments is constant change and evolution.

---

2 When we use the jigsaw metaphor, we often get pushback from fellow engineering leaders who proclaim that the correct metaphor is Jenga.

Many leaders, especially those from larger organizations, need help in chaotic environments because they expect the same support structures—like a prioritization framework or a project management methodology—to be in place. In a chaotic environment, you're not just working within these structures but building them.

We need to ensure our processes are:

- Flexible enough to accommodate change
- Strong enough to protect against fluctuating priorities

This balancing act is often a lot harder than it looks. Later in the chapter, we discuss some techniques to help you strike a balance between these, ensuring that you execute and deliver what is required of you and your team, rather than being mired in an endless bog of changed priorities.

Prioritization Techniques

We've established that in chaotic and rapidly changing environments, you must prioritize and balance your work to ensure that you're working on the right things in the right combination and sequence. Getting this right helps engineering leaders make informed decisions and deliver your organization's needs. Now, let's delve into the approach we propose for prioritizing software and product development and how it can help you balance work effectively.

First, consider some of the inputs you should evaluate when prioritizing tasks.

Key Inputs for Decision Making

Making good prioritization decisions does require some base information—not perfect information, but some initial understanding of why you are doing the work, its impact, and a broad understanding of any requirements. More information is better, but this isn't always available or complete. When it comes to prioritizing work, several critical factors should guide your decision-making process:

Business impact
> Evaluate how each task contributes to the product's or organization's goals, or both. Does it drive revenue, enhance customer satisfaction, or open new market opportunities? Tasks with a higher business impact typically deserve higher priority.

Urgency
> Consider each task's time sensitivity. Are there deadlines imposed by go-to-market needs, contractual obligations, or competitive pressures? Urgent tasks need to be addressed promptly to avoid negative consequences.

Effort
> How much effort is required to complete the task. Is it days, weeks, or months? You need to understand if a task requires more capacity than you have, or if its timeline will create issues that clash with the business impact or urgency.

People
> Assess your team members' availability and skill sets. Do you have the people to execute the tasks efficiently, and do those people have the right skills?

Dependencies
> Identify interdependencies between tasks. Some tasks can begin only after others are completed. This is common sense in most cases, especially for product development. However, other projects—such as moving an office or a complex deployment—may require a clearer understanding of the sequence of tasks.

We don't generally create separate systems for storing this information; we use existing tools and frameworks that include this scope in their remit.

Utilize Established Frameworks

Just as we consider agile a broad foundation for guiding how we operate, we also leverage established frameworks—in part or whole—when choosing how to prioritize work. These tools provide structured approaches to assess and rank tasks or features based on specific criteria. There are a lot of existing frameworks that you can draw upon to bootstrap prioritization, and many frameworks are already supported in tools like ticketing systems.

Like agile, we're not dogmatic about our choices or usage of these frameworks; we take the bits that work—sometimes from several frameworks—and shape them into a system that works for us and meets our objectives and constraints, especially the need to be flexible and adaptable.

As the base of our approach, we start with one of the classic techniques for understanding work: the user story.

You should find a tool to manage your prioritization, workload, and backlog. There are a lot of potential tools out there that can help, ranging from spreadsheets to Trello, right up to complex Agile management tools like Linear, Shortcut, and Jira (and a dozen others). We don't have a strong opinion on choosing a specific tool. Try a few out and see what works for you and your team. Different tools also have different audiences. A tool like Jira, for example, has an extensive collection of features and integrations that may be overwhelming to a small team (and hard to manage), so a more straightforward tool might be the better choice. Additionally, you might evolve through several iterations of tools as your needs change.

User stories

User stories center around mapping the user journey to visualize features based on user needs. It's a straightforward concept: imagine (or ask the customer) how they will use your product or a specific feature. The user story approach focuses on delivering an MVP that satisfies core user requirements, enabling you to provide value early on. While user stories are more of a product decomposition method rather than a prioritization technique in their own right, we think your prioritization technique needs to have inputs that maximize your ability to prioritize effectively. User stories are those inputs. Good user stories also encourage efficient use of your people, focusing on high-impact features (more on this shortly). They also encourage addressing high-risk items earlier, potentially reducing future issues. Finally, aligning development with customer needs often correlates with your business objectives.

We like user stories because they encourage and reinforce iterative development and continuous product refinement based on feedback and (optimally) data-driven insights. It emphasizes understanding the user's perspective, ensuring that development efforts align with user expectations and needs. You've likely seen user stories applied to product development—for example, "you're building a customer-facing product." However, we think they equally apply to other purposes, including internal tools and platforms. Like customer-facing products, those areas have users too.

In Practice: April's Story

At PixelCurl, April recognized that user stories aren't just for customer-facing features. To address the mentorship gap, she created a user story for an internal tool: "As a junior engineer in StudioOps, I want to easily find and book time with senior engineers who have expertise in specific areas, so that I can get unblocked quickly and learn from their experience." This frames the mentorship problem as a solvable product challenge rather than an abstract cultural issue.

Typically, user stories are written by product managers or product owners; though in practice, developers or others can also write them. The point is to create something that influences functionality from a user's perspective. Some teams rely on the product owner to craft them and shape the product backlog, while others let anyone contribute. The source doesn't really matter as much as making sure the story reflects real-life user goals.

User stories often follow a known template that helps keep things straightforward. You've probably seen something like the format shown in Figure 6-1: as a [role], I want [capability], so that [benefit].

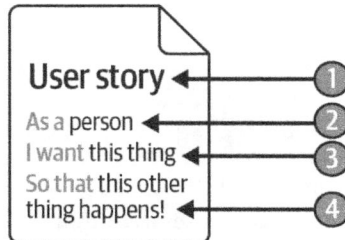

Figure 6-1. The user story pattern

There are variations, too. Some emphasize the benefit first, like: in order to [receive benefit] as a [role], I can [goal/desire].

Others add more context (the who, when, where, what, and why), or even consider malicious scenarios with "evil user stories" so that the team thinks like a hacker—for example, "As a disgruntled employee, I want to wipe out the production database to hurt the company."

User stories can be just a few sentences long. They're short and easy to reason about, which helps you stay focused on user value rather than drowning in specs. For example, imagine you're working on a software as a service (SaaS) tool and thinking about adding two-factor authentication. The user story could be:

> As a registered user, I want to enable two-factor authentication using an authenticator app or via text message to secure my account with an additional verification step beyond my password.

You can see we've identified registered users as our audience. We've also made our request: develop two-factor authentication that supports an authenticator app or a text message as the second factor. Finally, we've identified why the user wants this feature. That's it. It's very simple, compact, and easy to understand. You don't need to include implementation details in a user story; the home for those is probably a request for comment (RFC), ticket, or commit message/pull request.

Acceptance criteria. Teams use acceptance criteria to ensure a user story is truly complete. These criteria define what the product owner needs to say: "Yes, this meets our requirements." The criteria help set boundaries around a story, making it clear when the work is done. Some teams like the given-when-then format (which can be expressed as preconditions, actions, and results) for acceptance tests. We can see this in the two-factor authentication acceptance criteria for our prior user story.

Using the given-when-then acceptance criteria, this would look as follows:

Given

> The user is registered, logged in, and on the account security settings page.

When

> They select the option to enable two-factor authentication and choose "Use authenticator app."

Then

> Two-factor authentication should be successfully enabled.

Instead of this framework, others just use a list of items, but the key is that every piece of acceptance criteria must be met.

A start, not a solution

That said, user stories aren't perfect. Firstly, their simplicity means that they don't scale well. I've seen plenty of teams meet a block when their user stories start multiplying like rabbits. It's like trying to juggle while riding a unicycle; you're managing an ever-growing pile of stories while personal opinions and biases keep you off balance. And let's be honest, we've all been there with those deceptively simple user stories that turn out to have more gaps than Swiss cheese. By design, stories are open to interpretation and usually don't spell out every last detail. They're conversation starters, not contracts.

To manage this complexity, many teams group stories into bigger chunks. Large collections of stories that focus on a common goal are generally called "epics." Multiple epics, or large groups of related stories, can form "themes," and in some scaling frameworks, groups of themes or epics might roll up into "initiatives" or "programs." Agile management tools encourage you to think of work hierarchically: user stories roll up into epics, epics roll up into initiatives, and so on. But remember, these terms vary and can mean different things depending on who you talk to. There's no universal rule—just what works for you and your team.

Nonfunctional requirements, like performance or security standards, can also be overlooked. Sometimes, technical complexity doesn't neatly map to a single story focused on user-facing value. You might need "technical-only," "architecture-only," or "infrastructure-only" stories.

The real headache comes when your project needs to change direction quickly. Those carefully structured user stories can start feeling too restrictive—like concrete shoes when you need to dance quickly—and before you know it, you're so focused on checking boxes that you might miss the next big opportunity coming around the corner. It's a bit like being so focused on following your GPS that you miss the amazing shortcut the locals all know about.

Despite their limitations, we still rely on user stories because they emphasize the right things: user value, early delivery of workable features, and continuous improvement. Of course, as a project grows more complex or needs to pivot quickly, you will find yourself juggling a growing number of stories and needing help to keep them all aligned. At this stage, you can start to mix and match from other established frameworks, taking the bits that work for you now and providing a space and opportunity to grow into a larger framework as you scale.

Impact-versus-effort analysis

To help mitigate the downsides of user stories, like when you're faced with a mountain of them and need a more structured way to decide what comes first, we layer on top a prioritization framework. We like the very simple, but powerful, impact-versus-effort matrix. This framework complements user stories by introducing a method that helps you decide which stories (epics, features, tasks) to tackle next.

The impact-versus-effort method is straightforward:

Impact
How much potential benefit (sometimes also described as impact) does implementing this feature or completing this piece of work provide to users, customers, or the business? Does it align closely with strategic goals? Will it improve user satisfaction, increase revenue, or open new market segments? High-impact items significantly improve user experience, enable revenue growth, or solve an existing pain point. Low-impact features are often nice-to-haves, aesthetic refinements, or changes that yield minimal tangible benefit.

Effort
How much time, skill, and complexity are required to implement this feature or perform this task? Is it a quick fix or a long, involved project? Will it tie up your team's most skilled engineers for weeks, or can a single developer handle it efficiently? High-effort items might need complex architecture changes, development time, or scarce specialist expertise. Low-effort tasks are quick to implement, require minimal resources, and don't introduce much risk.

You can quickly see where items fall by plotting features or user stories on a two-dimensional grid, as in Figure 6-2. Impact is on one axis and effort is on the other.

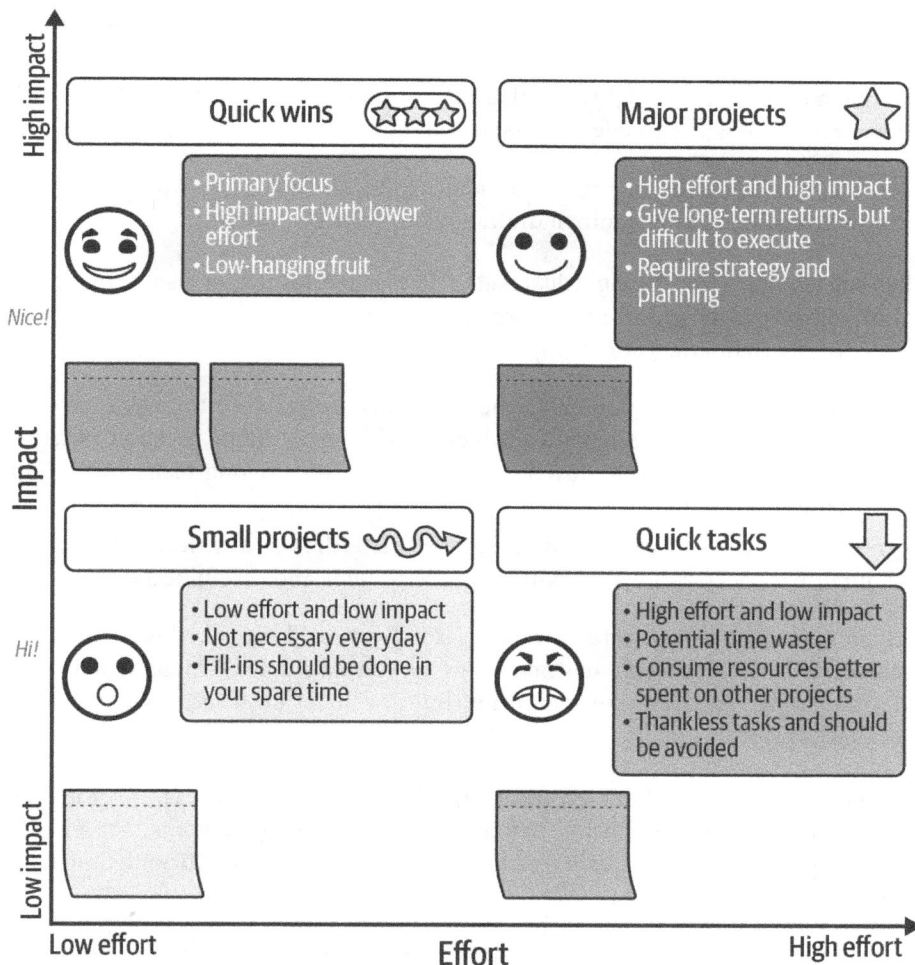

Figure 6-2. Impact-versus-effort quadrants

Plotting user stories or features on a grid with impact on one axis (vertical) and effort on the other (horizontal) creates four logical quadrants. Each quadrant guides how you think about prioritizing the work.

1. High impact, low effort (top-left quadrant). This quadrant is the "sweet spot." These features deliver significant benefits for minimal investment. They're often low-hanging fruit and can quickly boost user satisfaction or unlock meaningful improvements. For example:

- Adding a simple filter to a product listing's page when users have asked for it for ages
- Implementing a minor UI tweak that makes navigation more intuitive, increasing user retention with just a few hours of coding

Features in this quadrant are prime candidates to tackle first, as they provide a strong return on investment with minimal drain on your team's resources.

2. High impact, high effort (top-right quadrant). These features promise substantial benefits but come with a significant resource cost. They might still be worth doing, just not necessarily right away. For example:

- Rebuilding a core component of the platform's architecture to enable new scalability or performance benefits, which might open doors to large enterprise customers or dramatically lower operating costs in the long term, but requires several weeks of engineering time and careful planning
- Introducing a brand-new product module that addresses a critical customer need but requires deep integration with multiple systems and careful testing

Work in this quadrant is often strategic. You'll probably break down these large initiatives into smaller steps, or time them so you can deliver incremental impact while mitigating risks. They're important, but you must be smart about when and how to tackle them.

3. Low impact, low effort (bottom-left quadrant). These features are easy to implement but don't move the needle much. They might be minor improvements that a handful of users requested or small "quality-of-life" changes that don't significantly impact key metrics. Examples include:

- Changing the background color of a page or adding a minor customization option that only a few users have mentioned
- Tweaking a tooltip or copy that doesn't alter user behavior or business outcomes but is quick to fix

Because these tasks are low-hanging fruit in terms of effort, you may pick them up when you have a spare moment or as a quick morale booster, but don't let them dominate your roadmap. They shouldn't displace higher-impact work.

4. Low impact, high effort (bottom-right quadrant). This is the "danger zone" for prioritization. These features require significant time and resources but offer little corresponding benefit. They might come from vague stakeholder requests that lack a solid business case or "pet projects" that aren't strongly connected to user needs. Some examples might be:

- A complex, behind-the-scenes refactoring project that makes the codebase slightly cleaner but doesn't impact performance or reduce maintenance costs measurably
- Implementing a feature no one specifically requested, which requires intricate design and multiple development sprints but doesn't align with strategic goals

Generally, you deprioritize or avoid these features. If one keeps coming up, challenge its necessity or see if it can be reimagined in a more valuable, lower-effort form.

In the sidebar, consider how much April might make use of the matrix.

In Practice: April's Story

April uses this framework to prioritize her competing initiatives at PixelCurl. Creating a comprehensive onboarding program (high impact, high effort) would significantly improve new hire integration but requires weeks of development. Meanwhile, instituting weekly cross-team code reviews (high impact, low effort) can immediately start breaking down silos with minimal setup. She tackles the code reviews first while planning the onboarding program for the next quarter, demonstrating to leadership that she can deliver quick wins while working toward strategic goals.

The impact-versus-effort matrix is a lens to help you see where to invest your team's limited time and energy. High-impact, low-effort items are usually first picks, offering quick wins. High-impact, high-effort tasks still matter but may need careful staging or breaking down into smaller, more manageable pieces. Low-impact, low-effort items can fill gaps in your schedule but shouldn't crowd out more impactful work. Finally, low-impact, high-effort features should be rigorously questioned and often deferred or scrapped entirely.

Understanding these four quadrants gives you a roadmap for turning a random assortment of user stories, features, and ideas into a coherent, strategic plan that balances immediate impact with long-term goals. This helps mitigate the downsides of user stories, their tendency to proliferate and become challenging to manage, by giving you a clear, visual framework to prioritize what matters most.

Impact-versus-effort analysis helps cut through the noise. Instead of wading through a long backlog of user stories all with seemingly equal priority, you have a tool to assess their relative worth. A story that seems promising but requires a substantial architectural overhaul (high effort) might not outrank a simpler feature that delivers immediate user benefit (low effort, high impact). Similarly, you can deprioritize those features that would chew up a lot of resources without moving the needle on user happiness or business objectives.

This approach does more than just help you pick what to do first; it also invites conversation. When the team sees that a particular story ranks low in impact but high in effort, it encourages exploring alternatives—maybe a simpler solution achieves a similar outcome. Or perhaps a story that initially looked low impact becomes critical as business circumstances evolve. Impact-versus-effort analysis isn't static; it's a framework you revisit as priorities shift and new insights emerge.

By layering the impact-versus-effort analysis over user stories, you bring more structure to your decision-making process without losing the flexibility and user-centered thinking that user stories provide. It helps you avoid getting bogged down in an ever-growing backlog, keeps you focused on delivering meaningful outcomes, and encourages iterative decision making. As a result, you can more confidently say "yes" to the work that matters and "not now" to the distractions that might otherwise dilute your team's efforts.

Ultimately, this pairing—user stories at the core for clarity and empathy with your users, and the impact-versus-effort analysis for pragmatic prioritization—forms a balanced toolkit. It lets you maintain agility while ensuring that the features you invest in will most likely deliver real, tangible impact in a chaotic and rapidly evolving environment.

This framework process also scales amazingly well. It can be applied to everything from a daily to-do list, to a product build, to large-scale organization-wide planning.

Finally, combining user stories and a prioritization framework leaves you with a pool of work to manage: a backlog.

Exercise: The Impact-Versus-Effort Mapping Session

Practice prioritizing your current chaos using the impact-versus-effort framework.

Step 1: List 10–15 items currently competing for your team's attention (features, bugs, technical debt, team initiatives, operational tasks).

Step 2: Write each as a simple user story if applicable, or a clear one-sentence description if not.

Step 3: Plot each item on a 2 × 2 grid with impact (vertical) and effort (horizontal) axes. Use relative sizing: don't overthink exact measurements. For each quadrant, answer:

- High impact, low effort: which two to three items will you tackle this week?
- High impact, high effort: which one needs to be broken down into smaller chunks?

- Low impact, low effort: which items are "Friday afternoon" tasks when you have spare cycles?
- Low impact, high effort: which items will you officially deprioritize or remove?

Step 4: Share your grid with a colleague or your team. Where do you disagree? These disagreements are valuable data about differing perspectives on impact.

Managing Backlogs Effectively

An organized, thoughtfully curated backlog is at the heart of effective product management. But I hear you saying, "I am not a product manager; product managers manage backlogs." Yep, it's true that they often do. But one of the many charms of chaotic organizations, that we described in Chapter 1, is that ownership is often diffuse or unclear. Both Juan and James have found themselves managing backlogs—either with the product managers who own them or by acting as de facto product managers to fill gaps or address a lack of ownership. Knowing how to do this helps you and your team manage your work.

When your backlog accurately reflects current priorities, the team can confidently focus on the tasks that matter most. To ensure this, commit to regular review sessions that fit in with your development lifecycle—for example, weekly or biweekly if you have one- or two-week sprints. In these sessions you will reassess items and confirm if they're still relevant and aligned with strategic objectives. These sessions help you spot outdated user stories or features that were once critical but no longer support evolving business goals. Over time, pruning these stale items keeps the backlog lean, preventing it from becoming a confusing inventory of long-forgotten ideas.

We prefer to prune (delete) stale items, keeping the following in mind:

- You already have more work to do than capacity to work on it.
- If it's important, it will come up again.

But if you're concerned about pruning too aggressively—as not everyone can be a ruthless pruner—then introduce an "icebox" or "ideas" category to maintain clarity. Instead of cluttering your active queue with suggestions that lack immediate impact, park them in the icebox. This is a space for those intriguing concepts that might be useful in a different market phase, a later product iteration, or after specific dependencies are resolved. By stashing potential work here, you preserve good ideas for future exploration without burdening the current workflow.

Above all, don't maintain the backlog alone. Implement collaborative backlog grooming sessions, where product managers, developers, designers, and other stakeholders review and refine items and offer diverse perspectives and insights. This collective curation process fosters a sense of shared ownership and ensures everyone understands the "why" behind each entry.

In Practice: April's Story

April transformed PixelCurl's backlog grooming sessions from technical-only discussions to collaborative planning meetings. She ensures Jordan and Priya, her engineering managers, actively participate in prioritizing not just features but team development initiatives. When a senior engineer suggests deprioritizing junior training in favor of a new feature, the group discusses the long-term velocity impact of under-mentored developers, making the trade-off visible to everyone.

These discussions often surface nuanced details—like a hidden dependency or an unexpectedly simple workaround—that refine the feasibility and desirability of particular tasks. The result is a backlog consistently aligned with real-world constraints, customer needs, and strategic priorities, ultimately enabling the team to deliver tangible outcomes.

A Good Cadence

Our experience has been that establishing a cadence of activities to manage your work and backlog is crucial. A cadence not only enforces operational management of your work but also provides structure and consistency for your team. Finding the cadence that works for you might require experimenting, trial, and testing. We usually start with Agile-style sprints.

You're probably familiar with sprints: short, time-boxed periods during which a team focuses on completing a set of user stories. We generally find that one-week sprints are too short, but a month is too long, especially in chaotic environments where changes in people and priorities can be frequent. So, we settle for a middle ground of two-week sprints. We then set a schedule within the sprint. Like most Agile-esque systems, we like to start sprints with planning. For example, a cadence we often use is as follows:

Week 1
   On Monday, we conduct sprint planning. From Tuesday through Thursday, we hold morning standups, either in person over virtual meetings, or on Slack. On Friday, we have asynchronous standups via Slack or another shared platform.

Week 2
   We again hold daily standups on Monday through Thursday. On Friday, we conduct demos and a retrospective on the sprint.

This cadence enables the team to plan which items from the backlog can be realistically finished. Then, the team proceeds with daily check-ins to ensure everyone remains on the same page and can quickly address impediments. At the end of the sprint, the team conducts a review to demonstrate their accomplishments and a retrospective to discuss successes, challenges, and opportunities for improvement.

This cadence is just an example of a myriad of combinations of cadences that exist in the Agile world. If you find daily standups are too frequent, then try every second day. The important lesson here is to see what works for you, and then to keep testing and confirming that it continues to. If not, adjust accordingly.

Approaching Estimates Wisely

In the previous section, we quickly moved over the phrase "realistically finished." But how do we decide what work will fit into our cadence? Well, we determine this with estimates. This part of the chapter is where the classic refrain of "Danger, Will Robinson!" echoes in our heads. Estimates are both powerful and dangerous. They are an important, if imperfect, compass guiding your team toward realistic planning. Yet the quest for hyper-accurate estimates often devolves into contentious debates that waste time and erode trust.

Instead of seeking perfection, focus on consistency. Choose a method, story points, time-based estimates, or a more whimsical approach like assigning "difficulty tokens," and ensure everyone uses it in the same spirit. What matters most is that each team member similarly interprets the scale. Over time, these consistent standards enable you to compare tasks with reasonable confidence, making it easier to identify what's genuinely complex versus what just seemed tricky at first glance.

Avoid Bikeshedding on Estimations

Beware of the tendency to spend disproportionate time debating an estimate or, worse, debating the best estimation method. While agreeing on a shared framework is good, you don't want to squander valuable energy on this secondary concern. Instead, acknowledge that any method will have its shortcomings and move forward, refining your approach as you gain more experience and data. Embrace approximation and uncertainty. Early in a project, you may accept broad estimate ranges—such as ±25%—and then narrow these ranges as tasks become clearer or you learn from past sprints. This openness to uncertainty reduces pressure on the team and sets realistic expectations for stakeholders.

Above all, keep communication transparent. Say so if a feature's timeline is uncertain due to technical complexities or unknown dependencies. Stakeholders are more likely to remain supportive when they understand that estimates are guideposts rather than guarantees. When everyone recognizes that estimates are tools for informed decision making and not ironclad promises, the team can move forward more confidently and make adjustments as and when new information emerges.

Chapter 10 discusses estimates in the context of measurements, metrics and, importantly, measuring velocity.

Balancing Different Types of Work

Up until now, we've focused on building features or new products. A genuinely resilient development process involves more than just churning out new features. While shiny capabilities and innovative enhancements draw attention from customers and shareholders, they are only one dimension of long-term success. Behind every successful product is a delicate equilibrium: you must also allocate time to fix bugs that degrade user experience, address technical debt that inhibits future scalability, and perform routine maintenance tasks that preserve system stability.

It can be helpful to imagine your product as a living ecosystem. If you focus exclusively on new feature development, then your codebase might become fragile, with performance or architectural flaws lurking beneath the surface. These flaws eventually slow everything down and threaten the product's reliability. Conversely, if you devote too much effort to technical debt and maintenance, you risk stagnation—missing opportunities to differentiate your product and delight users with new functionality. By striking a balance, you ensure that your team's efforts work together. Stability supports innovation, and innovation justifies the investment in ongoing caretaking.

A lot of people look at bugs and tech debt as separate sets of tasks, sometimes even separate backlogs. A more practical way to maintain this balance is to include all types of work—new features, technical debt reduction, bug fixes, and maintenance—in the planning and prioritization discussions and in one backlog. Treat each category as a priority in your backlog. By making all these streams visible, measurable, and subject to the same prioritization criteria, you naturally create a healthier distribution of efforts. This approach leads to a more adaptable codebase and a product that can meet user needs without succumbing to the gravitational pull of hidden issues.

In the sidebar, consider how April takes advantage of this approach to a backlog.

Addressing Technical Debt

Technical debt is a topic worth covering in its own right. Technical debt is often invisible until it's not. It accumulates like unseen rust, quietly corroding the flexibility and efficiency of your codebase. For nonengineering stakeholders, the concept can be abstract, so it's essential to translate it into concrete terms. Explain how certain short-cuts that saved time in the past now cause longer development cycles, complicate integrations, or create stability risks. Frame technical debt as a tax you're continually paying, either in slower release cycles or increased defect rates. When people see how this "invisible burden" affects the bottom line and the product's long-term health, they become more supportive of investing in its reduction.

Quantifying technical debt helps solidify its importance. For instance, show stakeholders that resolving a piece of technical debt could cut the time needed to deliver future features by 20%, or highlight how addressing a brittle subsystem reduces the likelihood of critical outages. With metrics in hand, you can integrate technical debt reduction into your regular planning cycles, scheduling dedicated sprints or time blocks to tackle these issues. Just like new features, technical debt items should be prioritized based on their impact. So fix the most obstructive problems first, and then move on to less critical ones.

A technique we've adopted from product engineering is the user story for tech debt. A lot of people keep tech debt as tickets—for example:

> "Our React pages and components are duplicative and unwieldy and need to be refactored."

This tells you exactly the problem but nothing about why it matters and what a good outcome might look like. Instead, we could convert it into a user story:

> As a developer maintaining our product's frontend, I want to refactor our duplicative and unwieldy React pages and components to make them simpler, reusable, and easier to integrate. This will help us deliver new features more quickly, reduce defects caused by tangled code, and ultimately ensure a more stable and maintainable application over time.

Add in prioritization, and now you're a step closer to comparing the apples-to-apples value between this technical debt and a new feature and balancing your feature development with tech debt.

Taking a proactive stance prevents technical debt from spiraling out of control. Instead of letting it accumulate until a crisis occurs, incorporate routine cleanups that make your codebase healthier. Over time, this will make your engineering effort more predictable, your product more stable, and your team more confident. By persistently paying off the debt, you will invest in your product's future agility and resilience.

Exercise: The Process-Debt Audit

Identify and articulate "organizational debt" that's slowing your team down.

Step 1: Set a timer for 15 minutes. List every process friction point, communication breakdown, or organizational inefficiency currently affecting your team. Don't filter; just brain dump.

Step 2: Choose your top three most painful items and convert them into user stories using this format: "As a [role], I want [capability], so that [benefit]."

Step 3: Add acceptance criteria for each of the three top items. For each story, identify the following:

- Current cost (hours lost per week, morale impact, delivery delays)
- Minimum viable fix (the 20% effort that gets 80% improvement)
- Success metric (how you'll know it's better)

You can then add these to your actual backlog alongside technical work. If that feels uncomfortable, ask yourself: would you tolerate the equivalent technical debt?

Monitoring Progress and Adjusting Priorities

Maintaining a sense of direction in chaotic, fast-moving environments can be challenging. This is why regular progress tracking is indispensable. By defining clear success metrics for various categories of work, like the number of successfully shipped features in a quarter, the average resolution time for critical bugs, or the reduction in certain performance bottlenecks, you create a basis for informed decision making. These metrics act as beacons, helping you see whether your team's current trajectory aligns with larger organizational goals.

We'll talk more about metrics specifics in Chapter 10.

Holding periodic reviews weekly, biweekly, or monthly, enables you to step back and evaluate whether the chosen priorities pay dividends. If a certain initiative isn't yielding the expected results, it might be time to pivot. If an unforeseen opportunity arises, you might reallocate resources to seize it. This flexibility ensures you're not locked into a static plan that doesn't respond to reality. Instead, you can gracefully shift gears as conditions evolve, ensuring your efforts remain nimble and outcome driven.

This adaptive approach also reduces wasted effort. Regularly assessing performance and outcomes minimizes the risk of pouring time into initiatives that no longer make sense. Instead of discovering misalignments after weeks or months of labor, you promptly identify them and correct course. The result is a more efficient use of resources and a team that feels empowered to navigate uncertainty rather than be hindered by it.

Communicating Priorities Transparently

Just reviewing and updating your priorities is the first step. Once you've done it, you need open and honest communication about the outcomes—even if the news isn't good. We love to shout our good news, especially in chaotic environments; a win is a win. But the bad news we tend to want to mumble or send via email on a Friday afternoon at 6:00 p.m. Honest and transparent communication helps transform chaotic conditions into manageable challenges. When you share not just what you're prioritizing but also why, you foster an environment of trust and understanding. For example, explaining that a particular feature leaped ahead in the queue because it directly addresses a top customer complaint gives context to your team and stakeholders. It shows that decisions aren't made arbitrarily; they're guided by a logic rooted in user feedback, business objectives, or technical constraints. The same is true of changed priorities; if you made a commitment, then transparently sharing the new situation might be painful, but nothing erodes trust faster than silently missed deadlines.

Encouraging open dialogue also enriches the decision-making process. Invite feedback from engineers who might know a quicker path to the same outcome or designers who can spot a user-experience improvement that boosts feature adoption. These conversations often uncover valuable insights that a top-down directive would miss. They also build a sense of community and shared purpose. Team members who understand the reasoning behind decisions are more likely to commit to them fully, lending their creativity and expertise to make the chosen path successful.

Regular updates—through sprint reviews, monthly check-ins, or informal Slack announcements—keep everyone in the loop about shifting priorities, celebrate successes, and share new challenges on the horizon. With visibility into changes and their rationales, team members adapt more readily, and stakeholders have fewer surprises. Over time, this transparent communication culture helps the team internalize the organization's values and objectives, reinforcing alignment even amid perpetual change.

By adopting these practices, you create a resilient operational framework: thoughtfully managing backlogs, estimating consistently yet flexibly, balancing various work streams, addressing technical debt head on, monitoring progress openly, and communicating priorities transparently. This framework empowers the team to navigate chaos confidently, adapt to new conditions, and continue delivering value in an environment where uncertainty is the rule rather than the exception.

Not Everything Is a Product

In the midst of prioritizing user stories and managing product backlogs, it's easy to forget that engineering teams often tackle work that doesn't fit the iterative product model. Moving offices, implementing compliance requirements, migrating infrastructure, coordinating vendor evaluations—these initiatives all have fixed endpoints, rigid dependencies, and linear sequences that don't benefit from constant iteration. Recognizing when to shift your prioritization approach is as important as having one in the first place.

We're not saying you should abandon everything we've discussed. But it's important to acknowledge that different types of work will demand different prioritization frameworks. The impact-versus-effort matrix still applies, but the execution model changes. When you can't A/B test an office move or iterate on a compliance deadline, forcing Agile methodologies onto fundamentally linear work creates friction rather than flow.

When Linear Makes Sense

Some work follows a logical sequence by necessity, not by choice. You can't install network infrastructure before signing a lease. You can't complete a compliance audit before gathering the required documentation. These projects have the following characteristics that make linear planning more effective:

Fixed dependencies
> Tasks must happen in a specific order, with little room for parallel work or reordering based on new information.

External constraints
> Vendor availability, regulatory deadlines, or physical logistics dictate timing and sequence.

Binary outcomes
> Unlike features that can be partially shipped or rolled back, these projects either succeed completely or fail entirely.

Limited iteration potential
> Once you've moved offices, you're not doing it again next sprint, based on user feedback.

In Practice: April's Story

For April, this distinction became clear when she needed to coordinate both a team restructuring (iterative and feedback driven) and a security audit (linear and deadline driven). April realized that trying to run the audit like a series of sprints would be as ineffective as trying to restructure the team with a rigid Gantt chart.

Creating Just Enough Structure

The key insight from earlier in this chapter bears repeating: the choice of planning process matters less than actually working with it. As Mike Tyson famously said, "Everyone has a plan until they get punched in the mouth." Your plan for linear projects needs to be robust enough to provide direction but flexible enough to survive contact with reality.

Start with the bare minimum structure needed to coordinate the work:

Define the endpoint clearly.
Unlike iterative product work where the definition of "done" evolves, linear projects need a clear finish line.

Map critical dependencies.
Identify what must happen before what else can begin. This isn't about creating elaborate charts; it's about understanding the nonnegotiable sequence.

Assign single owners.
Every task needs one person accountable for its completion. If a task needs multiple owners, decompose it further.

Build in checkpoints.
Even linear projects benefit from regular progress reviews. These aren't retrospectives aimed at process improvement; they're reality checks to ensure you're still on track.

Communicate the plan widely.
Everyone affected needs to see the full picture, understand their role, and know how delays in one area impact others.

Working the Plan, Not Serving It

The danger with any structured approach is that the plan becomes more important than the outcome. You've seen this before: teams so committed to their Gantt chart that they ignore obvious improvements, or they are so wedded to meeting predetermined milestones that they miss critical issues.

Remember our earlier principle: you serve the outcome, not the process. A plan for moving offices that doesn't result in a successful move is a failed plan, regardless of how well you followed it. This means:

- Adjusting timelines when reality diverges from estimates
- Reassigning work when someone's availability changes
- Accepting that some assumptions will be wrong
- Treating the plan as a living document, not fixed

Connecting Linear and Iterative Work

The real skill lies not in choosing between linear and iterative approaches but in running both simultaneously. Your team might be executing sprints for feature development while following a linear plan for a datacenter migration. The prioritization challenge becomes balancing these parallel tracks:

- Reserve capacity in your sprints for linear project tasks.
- Make dependencies between the two visible to everyone.
- Adjust sprint commitments when linear project milestones approach.
- Use the same prioritization criteria (impact versus effort) even if the execution model differs.

This dual-track approach reflects the reality of engineering leadership in chaotic environments: you're simultaneously shipping products iteratively while executing infrastructure changes linearly, building team culture gradually while meeting fixed compliance deadlines, and experimenting with new technologies while maintaining existing systems on schedule.

The frameworks and approaches you choose matter less than your commitment to making them work. Whether you're running sprints, managing a waterfall project, or some hybrid of both, success comes from adapting the process to serve your outcomes rather than becoming enslaved to methodology. Keep the focus on execution and maintain flexibility in your approach. Remember, the best plan is the one that actually delivers results.

Conclusion: Driving to Execution

Success as an engineering leader isn't about perfection or precision; it's about execution—effectively balancing multiple responsibilities while keeping your team moving forward. Throughout this chapter, we've explored how to build just enough process to create forward momentum without drowning in methodology. From user stories that keep you grounded in customer impact to the impact-versus-effort matrix that cuts through endless debates, these tools share a common thread: they're designed to be wielded, not worshipped. The frameworks you choose matter far less than your ability to adapt them to your reality. Whether you're running two-week sprints for product development, managing a linear office move, or tackling technical debt that's been haunting your codebase, the principle remains the same: start small, measure what matters, iterate based on what you learn, and never let the process become more important than the outcome.

Chaotic environments won't wait for you to implement the perfect prioritization system. Your team needs direction now, not after you've read every Agile Manifesto and memorized every estimation technique. By embracing flexibility, implementing practical frameworks, and maintaining clear priorities, you can navigate these challenges while building your team's confidence and capability. The tools and techniques in this chapter aren't prescriptions; they're ingredients you can combine based on what your team needs today. Tomorrow, when circumstances change or challenges arise, you'll adjust again. That's not failure; that's leadership in action.

In Chapter 7, we'll look at some of the mechanics of running a team like budgets, costs, and buying decisions.

Budgeting, Costs, and Vendors

Budgeting is something that most of us never expected to think about as engineering leaders. Most engineering leaders perceive themselves as making technical decisions, and they see finance and budget decisions as "business problems." Well, solving those "business problems" is how we get the people and resources we need to deliver products and services to our customers (also in the business) and, significantly, help us understand how we get paid, plan for our future needs, and how the business functions. Budgeting is also an anchor in a chaotic environment. It dovetails into your planning and prioritization process to help you understand the resources needed to deliver on your plans.

Budgeting is not just crunching numbers (although I guarantee you that you'll see a lot of spreadsheets; there is nothing that accountants like more than spreadsheets). Budgeting is a strategic process that enables you and your organization to anticipate challenges, seize opportunities, and maintain competitive advantages. By systematically budgeting for people and resources, you, your team, and your organization can create a roadmap that bridges the gap between your current position and your desired future state.

In this chapter, we explain why budgeting is necessary and discuss the basics of budgeting and cost management (as the other side of budgeting is working out what you've overspent on). And we discuss the procurement of products and services and how to deal with vendors. It's important to note that neither of us is an accountant (nor a lawyer nor medical professional), so this is a layperson's guide to understanding the basics of budgeting and costs. This is not professional financial planning or legal advice, and you should always work on budgets and costs with someone who actually knows what they are doing, like your chief financial officer or finance team.

Why Budgeting Matters

Throughout this book, we've discussed that doing something is (almost) always better than doing nothing. This applies doubly to budgeting. You might think budgeting is just about controlling costs or allocating money, but it's the foundation of turning your grand visions into reality. Think of it as the bridge between "Wouldn't it be cool if we…" and "We actually did it!"

Strategic Alignment

Let's start with aligning spending with your mission: strategic alignment. This isn't just corporate speak for "ensuring everything lines up." It's about ensuring that your spending matches your stated priorities. If you say customer experience is your top priority, but your budget shows that all your resources are going to internal tools, well…you might want to rethink either your strategy or your spending.

In Practice: April's Story

At PixelCurl, April faced this misalignment head on when reviewing the StudioOps Q1 budget. Despite claiming "collaboration and mentorship" as top priorities after losing two junior engineers to frustration, the budget told a different story: $15,000 allocated to individual JetBrains licenses for each engineer, $5,000 for personal productivity tracking tools, but zero dollars for pair programming tools like Tuple or CodeTogether. The "team building" line item turned out to be a lonely $500 for "quarterly pizza." So, April reallocated: she cut the productivity tracking tools that nobody actually used, negotiated a team license for a pair programming tool ($8,000), instituted weekly team lunches ($600 a month), and set aside $10,000 for a proper code review tool.

We need to develop clear frameworks for deciding what's worth investing in. This isn't about creating complex scoring systems (but if that's your thing, go for it); it's about having honest conversations about what matters most. You'll need to balance your short-term needs (keeping the lights on) with your long-term dreams (becoming the next tech unicorn). And yes, you'll need to coordinate across different teams—finance can't live in a bubble, and neither can engineering.

Cost Management

Cost management isn't just about finding the cheapest option for everything (though your chief financial officer might wish it was). It's about understanding the actual relationship between spending and value. We need to get clever about how we classify and track costs. Fixed costs are like rent—they don't care if you had a good or bad month; they're showing up either way. Variable costs are more like your coffee habits—they scale with your activity level (and stress level, let's be honest).

Understanding the income side of the equation

While engineering leaders don't typically own revenue targets, understanding how your costs relate to income transforms you from a cost center to a strategic partner. This is where unit economics come into play, and they're simpler than they sound. There are some key concepts you'll hear a lot about in unit economics, especially if you sell products and services to others:

Cost of goods sold (COGS)
> COGS represents the direct costs of delivering your product or service. For a software as a service (SaaS) company, this includes infrastructure costs, third-party service fees, and support costs directly tied to serving customers. If you're running $50,000 monthly in Amazon Web Services (AWS) costs to serve 1,000 customers, that's $50 per customer in infrastructure COGS. This number matters because it directly impacts your gross margin: the percentage of revenue left after subtracting COGS.

Customer acquisition cost (CAC)
> CAC is what it costs to land a new customer. While marketing and sales often own most of this, engineering contributes through the features and performance that make the product sellable. If your sales team is spending $5,000 to acquire each enterprise customer because your product lacks key features that competitors have, that missing functionality has a real dollar impact.

Lifetime value (LTV)
> LTV represents the total revenue a customer generates over their relationship with you. Engineering has a direct impact on LTV through product quality, reliability, and feature development. A bug that causes 10% monthly churn doesn't just affect your bug metrics; if your average customer pays $1,000 per month and churns after 10 months instead of 12, you've lost $2,000 in LTV.

Gross margin
> Gross margin is where it all comes together: (revenue − COGS) ÷ revenue. This percentage tells you how much money you have to cover everything else: salaries, office space, and that fancy coffee machine. Engineering teams that understand gross margin make better decisions. Should you optimize that database query that's consuming 20% of your infrastructure costs? If you're at 70% gross margin and trying to reach 80%, that optimization might be more valuable than the new feature that your product manager wants.

Making unit economics work for prioritization

Once you understand these metrics, every technical decision becomes a business decision. That refactoring project that reduces infrastructure costs by 30% directly improves gross margin. The authentication feature that enables enterprise sales could double your LTV. The performance improvement that reduces support tickets lowers your cost to serve each customer.

Consider how this changes conversations around prioritization. Instead of arguing whether Feature A or Feature B is more important, you can discuss the following:

- Feature A reduces infrastructure costs by $10,000 a month, improving gross margin by 2%.
- Feature B enables enterprise-tier pricing, increasing average LTV from $10,000 to $25,000.

This framework also helps justify engineering investments. If you need to hire a new DevOps engineer, frame it as, "This hire will reduce our infrastructure COGS by 25% through optimization, improving gross margin from 72% to 76%, which adds $2 million to our valuation at standard SaaS multiples."

Many people think that cost management is about cutting costs, but it's actually about understanding how costs behave in your organization. Sometimes, spending more in one area can save you money overall. It's like buying a good pair of boots—they might cost more upfront, but they'll last longer than cheap ones.[1] We must balance immediate savings against long-term value, which requires tactical thinking and strategic vision. We'll discuss cost management in the dedicated section later in this chapter.

---

1 All credit for this metaphor to Terry Pratchett's Sam Vimes "Boots" theory of socioeconomic unfairness.

Risk Mitigation

If there's one thing we've learned, the universe loves a good plot twist. Risk mitigation through budgeting is your way of saying, "Nice try, universe, but we thought of that." This isn't just about having a rainy-day fund (though that's not a bad idea). It's about understanding what could go wrong and having a plan and the resources to deal with it.

We need to consider financial risks (e.g., What if our largest customer leaves?), operational risks (e.g., What if our primary database server fails?), and strategic risks (e.g., What if our competitor renders our product obsolete?). Your budget must include provisions for addressing these risks, whether through direct investment in solutions or maintaining contingency funds.

Performance Monitoring

Remember when we talked about measuring things? Budgeting gives us another lens for this. We need to track financial and operational metrics, but the key is that they must provide us with useful insights. Vanity metrics are like vanity muscles; they might look good, but they're not helping you win the race.

We want leading indicators (early warning signs) and lagging indicators (confirmation that what we did worked). Think of it as a weather forecast versus looking out the window; both are useful, but for different reasons.

Resource Optimization

Resource optimization sounds fancy, but it's about making smart choices with limited resources (and resources are always limited, even at the biggest companies). This isn't just about money; it's about people, technology, time, and all the other things that help you get stuff done.

If Chapter 6 taught us anything, it's that prioritization is the art of deciding what to do when you can't do everything. Budgeting is simply prioritization with dollar signs attached. The impact-versus-effort matrix we use to determine which user stories to tackle next can also scale up to budgeting decisions. Should we hire another engineer or invest in better tools? Which option sits in the high-value, low-effort quadrant of our organizational priorities?

Think of your budget as your backlog at the company level. Just as you manage a product backlog with user stories, acceptance criteria, and regular grooming sessions, your budget requires the same disciplined approach but with broader impact. When you estimate a user story at two weeks of engineering effort, you're implicitly allocating budget; those two weeks represent salary, infrastructure costs, and opportunity costs. When you decide to tackle technical debt instead of new features, you're making a resource optimization decision that affects both your sprint planning and your quarterly budget.

We need clear criteria for making trade-offs, and these criteria should mirror your prioritization framework. Remember how we categorize work into features, bugs, technical debt, and maintenance? Your budget needs the same balance. Allocating 100% of resources to new features while ignoring technical debt is like planning sprints without ever addressing your bug backlog; it might work for a sprint or two, but eventually, the accumulated debt will force your hand.

The estimation challenges we discussed in Chapter 6 also scale up to budgeting. Just as we accept ±25% variance in early sprint estimates, your budget estimates for new initiatives will carry similar uncertainty. If there is a contractor that you think you need for three months, you should budget for four months. Similarly, for the infrastructure upgrade that is estimated at $50,000, have a conversation about what happens at $65,000. The same principles apply: acknowledge uncertainty, communicate it transparently, and narrow the variance as you learn more.

In Practice: April's Story

At PixelCurl, April sees this connection clearly when she realizes that her team's inability to accurately estimate story points has a direct impact on her ability to budget for the coming quarter. If the team consistently underestimates effort by 30%, her headcount planning will be off by the same margin. She starts treating budget planning sessions as extended sprint planning, using the same impact-versus-effort discussions but adding dollar amounts to each axis.

These decisions need to consider both immediate needs and long-term implications. Should we upgrade our infrastructure now or wait until next quarter? Apply the same logic you use for technical debt: what's the interest rate on this particular debt? If waiting costs you more in inefficiency, incidents, or lost opportunities than the upgrade itself, the decision becomes clear. Your budget is your organization's way of keeping score on these prioritization decisions, ensuring that your resource allocation matches your stated priorities.

Stakeholder Management

Your budget isn't just a planning tool; it's a communication tool. It tells investors and leadership what you're prioritizing, shows employees where the company is heading, and demonstrates to customers that you're investing in their needs. Transparency in your budgeting process builds trust, but it needs the right kind of transparency. Nobody needs to know how much you spend on coffee filters, but they should know about significant investments in new technology or capabilities.

Innovation and Growth

Here's where budgeting gets exciting (yes, really). Innovation and growth initiatives need explicit funding—they rarely happen by accident. We must set aside resources for both incremental improvements (enhancing our existing products) and transformative innovation (creating entirely new products).

This is like planting a garden; some plants will give you tomatoes this summer, while others are trees that won't bear fruit for years. You need both, and your budget needs to reflect that.

Organizational Learning

Every budgeting cycle is an opportunity to learn and improve. We should capture insights about what worked, what didn't, and why. This isn't just about getting better at budgeting—it's about getting better at everything the budget supports.

Think of each budget cycle as an experiment, where some things will work better than expected, but others worse. The key is to learn from both and adjust accordingly.

Competitive Advantage

Finally, remember that good budgeting can give you a competitive edge. It helps you respond more quickly to market changes, invest in the right capabilities at the right time, and maintain the flexibility to seize opportunities as they arise.

Instead of just being about controlling spending, budgeting is about enabling success. It's a tool that helps turn your strategy into actual results, and while it might not be the most exciting part of your job, getting it right makes all the exciting parts possible.

Finally, remember that you can always add processes to your budgeting, but it's much harder to remove them. As we described in Chapter 6, when discussing agile practices: start simple, learn what works for your organization, and build from there. The goal is to create a budget that helps you achieve your goals while effectively managing risks and resources.

Understanding Cost Management

Before you can build a budget, you need to understand what you're spending and how you're spending it. So, let's talk about everyone's favorite topic: costs! Okay, maybe not everyone's favorite, but understanding costs is like having a good map when exploring new territory; it helps you know where you are, where you can go, and what obstacles might be in your way.

But the thing about costs is that they only tell half the story. To make truly strategic decisions, you need to understand how your costs relate to the value you're creating.

Think of your organization's cost structure as a complex Lego creation, but one that needs to generate more value than it consumes. Some pieces are firmly locked in place (your fixed costs), while others can be easily swapped or adjusted (your variable costs). And critically, each piece should contribute to building something valuable enough that customers will pay for it. Understanding how these pieces fit together and how they connect to revenue is crucial for making smart decisions about your organization's future.

Fixed Costs: The Foundation

Fixed costs include factors like:

- Your office
- Permanent staff salaries
- Insurance premiums
- Core technology infrastructure

We need to manage these costs carefully. Too many fixed costs can make you rigid and inflexible, but too few might mean you lack the necessary infrastructure to grow. It's a balancing act, and like most balancing acts, it's trickier than it looks.

Variable Costs: The Flex Players

Variable costs, like your coffee budget, go up when you're busy and go down when you're not. The most significant item for most tech teams is usage-based tech services, such as cloud services and SaaS. Managing variable costs involves striking a balance between efficiency and effectiveness. You could switch to the cheapest materials, but if your product falls apart, that's not a win.

Step Costs: The Level-Up Costs

It's worth highlighting step-up costs because not all costs grow linearly. Remember playing video games where you'd reach a new level and suddenly need better equipment? Step-up costs are like that. They jump up at certain thresholds. Moving to a higher tier of SaaS, opening a new office, or expanding your infrastructure are all step-up costs. Before jumping, you must be sure you're ready or have at least considered the possibility of that next level.

Who Pays?

We also need to consider who owns the costs. In larger companies, this is an intricate dance. If one department consumes a service that might have people and infrastructure costs, do you bill that department? These are generally called direct costs. Some of them, like people, are usually easy to categorize and typically reasonably easy to allocate.

On the other hand, there are indirect costs, such as shared costs. A single customer likely consumes a percentage of your infrastructure spend, so you need to determine that percentage and who consumes what.

Even if you don't bill a specific department or customer, it's essential to understand who uses your services and what that costs. When budget time comes around, you must show that your investments have been worthwhile and that further investment, either continued or increased, is worth it.

The Human Side of Cost Management

People often forget that cost management is as much about people as it is about numbers. Having the right mentality is key. You should build a culture that considers costs. For example, choosing a new architecture that requires purchasing a new tool or platform, or incurring a step-up cost, will have an impact on expenditure. Your senior engineers should be able to consider that, even if it's just an extension of capacity planning or performance estimation.

Remember: just like with any other aspect of running an organization, the goal isn't perfection; it's continuous improvement. Start by understanding your costs, making informed decisions, and learning from and adjusting to them. And yes, sometimes that means admitting that the fancy coffee machine for the office was not the most strategic investment (but try taking it away now!).

Creating and Managing a Budget

Let's discuss creating and managing a budget. Much like planning a road trip, when creating a budget, you need to know where you're starting from, where you want to go, and what resources you'll need. In this section, we'll break down how to create and manage a budget that works.

You may not have to do this alone. Many companies have finance teams that gather your input and turn it into actual financial plans. But even with a chief financial officer and a finance team, you are ultimately spending the budget, and you are responsible for ensuring that it gives you what you need and that you're using it to plan.

Starting with History

It can be helpful to think of historical data as your organization's diary. It tells you what happened, not what you think happened. We want to look back at least three years if possible (though in chaotic environments, sometimes yesterday feels like ancient history). We're looking for revenue patterns that might show if your business has seasons or key customer dependencies; for example, many ecommerce companies see significantly increased variable costs around Black Friday and the holiday period as customer use increases. We need to understand expense trends to know where the money goes and identify those weird one-time incidents that pop up, like that office renovation that went way over budget.

When diving into historical data, we need the "what" (numbers) and the "why" (context). When thinking about that growth spike last year, consider whether it was because of your brilliant strategy, or if your competitor accidentally set their warehouse on fire. Context matters, and understanding the story behind the numbers is crucial for making good decisions.

Categorizing Expenses

Now, we break down our expenses into categories that make sense. If you've got a previous year's budget, you already have an initial list to build on. Start with your people costs, as these are usually your most costly line item. Beyond just salaries (though those are big enough), we're talking about the whole package: base pay that keeps the lights on at everyone's home, variable compensation that rewards good performance, and benefits because health care isn't free (speaking from a United States perspective). There are also expenses like training, conferences, and books, because we want our people to get better at what they do. You probably also want to factor in travel and expenses if your team is remote or travels to visit customers or other offices. Finally, consider entertainment expenses: office parties, mini golf days, or whatever floats your team's boat.

Infrastructure is like your organization's skeleton; you need it to stand up straight. This includes all your technology costs (sadly, the Cloud isn't made of cloud or, more excitingly, candyfloss), physical space (if you still have an office), and equipment (from laptops to coffee machines).

All these services make modern business possible—from software subscriptions to key SaaS platforms ranging from customer relationship management tools to observability platforms, recruitment tools, professional services like lawyers and accountants, and specialized security and data analysis support.

Headcount Planning and Costs

When we talk about people costs being your biggest line item, let's get specific about what that really means. Headcount planning isn't just about salaries; it's about understanding the full cost of a human being in your organization and planning for the complex dynamics of team growth.

The real cost of a person

When you're budgeting for headcount, the salary is just the beginning. In our experience, the "fully loaded cost" of an employee typically runs 1.3–1.8 times their base salary, depending on your location and benefits package. Here's what you're actually paying for:[2]

Base salary
> The number everyone focuses on

Payroll taxes and government mandates
> Social Security, Medicare, and unemployment insurance (roughly 7.65% in the US, but this varies by country)

Benefits
> Health insurance, retirement contributions, life insurance (can be 20–30% of base salary)

Equipment
> Laptops, monitors, software licenses ($5,000–$10,000 annually for engineers)

Office space and utilities
> Even in remote setups, you might provide stipends

Training and development
> Conferences, courses, books ($2,000–$5,000 per person annually)

Recruiting costs
> Whether agency fees (15–25% of base salary) or internal recruiter time

Onboarding inefficiency
> Considering that new hires typically operate at 50% productivity for their first three months

---

2 These stats are based on recent data from the IRS (*https://oreil.ly/2uxPV*).

Planning for growth and attrition

Your headcount plan also needs to account for both growth and natural turnover. Industry average attrition for engineering teams runs 10–15% annually, though it can spike higher in hot job markets or drop lower in economic downturns. This means that for a 20-person team, you should budget to replace two or three people per year just to maintain current capacity.

When planning for growth, consider the rule of temporary productivity loss: every new hire temporarily reduces team productivity before increasing it. The team spends time interviewing, onboarding, mentoring, and context sharing. A good rule of thumb is that adding a new person reduces team capacity by 10% for their first month, 5% for their second month, and then becomes additive in the third month.

Timing your hires

The timing of headcount additions can make or break your budget. Consider the following:

Seasonal patterns
Hiring in Q4 is often harder and more expensive due to holidays and bonus cycles.

Batch versus continuous hiring
Hiring three engineers at once creates an onboarding bottleneck, but it might be more efficient for recruiter time.

Ramp time aligned with project needs
If you need someone productive for a project in Q3, you should be hiring in Q1

Budget cycle alignment
Many companies approve headcount annually; missing your window might mean waiting another year.

Seniority mix and team balance

Additionally, not all headcounts are created equal. A senior engineer might cost twice as much as a junior, but doesn't necessarily provide twice the value—sometimes more, sometimes less, depending on your needs. Consider your team's pyramid, as shown in Table 7-1. (These salary figures are estimates rather than hard and fast bands.)

Table 7-1. Estimated engineering salary ranges by seniority level with key role considerations

Engineering level	Cost estimate	Considerations
Senior	$180,000–$300,000+	Technical leadership, architecture, mentorship
Mid	$120,000–$180,000	Core execution, growing autonomy
Junior	$80,000–$120,000	High potential, need investment, future pipeline

The ideal ratio varies by company stage and needs. Startups might skew senior-heavy for speed, while established companies might invest more in juniors for long-term growth.

Hidden costs and failure scenarios

Budget for hiring failures because not every hire works out, and the cost of a bad hire can be two to three times their annual salary when you factor in opportunity cost, team disruption, and replacement costs. You should build in contingency for failed hires. For example, we typically budget 10–15% extra to account for the possibility of hiring mistakes, as well as extended search times. For example, key roles might take three to six months to fill. Can you afford to hire contractors in the meantime?

Candidates will often make counteroffers, or you may be competing against other opportunities. For this, you might need to go 10–20% above the initial budget to land critical talent. Finally, consider retention adjustments. Keeping your existing team might require mid-cycle adjustments, or if a key resource receives an offer, you may need to consider establishing a rainy-day fund for counteroffers.

The headcount business case

Finally, when requesting a headcount, always frame it as an investment, not a cost. Focus on the impact of the new employee. Will they improve revenue? You should be able to answer with, "This engineer will enable features that drive $X in revenue." There's also potentially an opportunity cost: "Delaying this hire means pushing the product launch by $X months." In some cases, you can present adding people as cost avoidance: "Without this DevOps hire, we'll spend $X on contractors and outages because we'll continue to have this unstable platform." Lastly, consider team health, morale, and retention. If you have people burnt out or on the verge of burnout, then adding headcount might not only provide capacity but stop loss. This could be justified by saying, "Current overtime rates are unsustainable and risk losing existing team members."

Remember that headcount is simultaneously your biggest expense and your greatest asset. Plan thoughtfully, budget comprehensively, and always consider the full lifecycle cost of a hire—from recruiter to resignation. The numbers might seem daunting, but understanding them helps you make informed decisions about when, who, and how to grow your team.

Setting Goals That Make Sense

We've discussed goal setting earlier in the book, and every goal or program you have defined will likely need a budget, either in terms of people or resources. We're not going to dive into goals in any depth (see Chapter 5 for that), but goals need to be backed by investments. Again, if you have goals that don't align with how you spend your budget, you must create that alignment before committing to things you can't pay for. Budget goals also need metrics; some goals will be absolute numbers (e.g., "reach $10 million in revenue"), while others might be relative (e.g., "grow faster than the market"). The key is to ensure that these goals align with your overall strategy and capabilities.

Remember: goals should be like good coffee—strong enough to get you moving, but not so strong that they give you anxiety. They should push you to improve while remaining achievable with focused effort.

Creating Your Budget Template

Your budget must be both detailed enough to be helpful and straightforward enough for people to use. It should show your costs (and potentially your income) and be flexible enough to update as things change. It is a living document that helps guide decisions rather than restrict movement.

Now we need to populate all of this into a spreadsheet. If you have one, your finance team might provide a template, but just in case, we're providing one in Figure 7-1 to get you started.

	Starting Balance	1	2	3	4	5	6	7	8	9	10	11	12	Total
Starting Balance	1,900,000													
Total Income		1,000	1,500	2,500	5,000	15,000	25,000	35,000	50,000	75,000	110,000	140,000	200,000	
Total Expenses		98,810	92,810	92,810	97,810	97,810	97,810	97,810	102,810	102,810	107,810	107,810	107,810	
NET (Income - Expenses)		(97,810)	(91,310)	(90,310)	(92,810)	(82,810)	(72,810)	(62,810)	(52,810)	(27,810)	2,190	32,190	92,190	
Projected End Balance		802,190	810,880	720,570	627,760	544,950	472,140	409,330	356,520	328,710	330,900	363,090	455,280	

INCOME

	1	2	3	4	5	6	7	8	9	10	11	12	Total
Sales	1,000	1,500	2,500	5,000	15,000	25,000	35,000	50,000	75,000	100,000	150,000	200,000	660,000
Subtotal	1,000	1,500	2,500	5,000	15,000	25,000	35,000	50,000	75,000	110,000	140,000	200,000	660,000
Total INCOME	1,000	1,500	2,500	5,000	15,000	25,000	35,000	50,000	75,000	110,000	140,000	200,000	660,000

EXPENSES

	1	2	3	4	5	6	7	8	9	10	11	12	Total
People													
Engineering Manager	16,000	16,000	16,000	16,000	16,000	16,000	16,000	16,000	16,000	16,000	16,000	16,000	192,000
Senior Engineer	15,000	15,000	15,000	15,000	15,000	15,000	15,000	15,000	15,000	15,000	15,000	15,000	180,000
Senior Engineer	15,000	15,000	15,000	15,000	15,000	15,000	15,000	15,000	15,000	15,000	15,000	15,000	180,000
Engineer	12,000	12,000	12,000	12,000	12,000	12,000	12,000	12,000	12,000	12,000	12,000	12,000	144,000
Engineer	12,000	12,000	12,000	12,000	12,000	12,000	12,000	12,000	12,000	12,000	12,000	12,000	144,000
Product Designer	14,000	14,000	14,000	14,000	14,000	14,000	14,000	14,000	14,000	14,000	14,000	14,000	168,000
Subtotal	84,000	84,000	84,000	84,000	84,000	84,000	84,000	84,000	84,000	84,000	84,000	84,000	1,008,000
Infrastructure													
AWS	5,000	5,000	5,000	10,000	10,000	10,000	10,000	15,000	15,000	20,000	20,000	20,000	145,000
New laptops	6,000	0	0	0	0	0	0	0	0	0	0	0	6,000
[itemized expense]													
[itemized expense]													
[itemized expense]													
[itemized expense]													
Subtotal	11,000	5,000	5,000	10,000	10,000	10,000	10,000	15,000	15,000	20,000	20,000	20,000	151,000
Services													
GitHub	600	600	600	600	600	600	600	600	600	600	600	600	7,200
Cloudflare	250	250	250	250	250	250	250	250	250	250	250	250	3,000
Sentry	100	100	100	100	100	100	100	100	100	100	100	100	1,200
PagerDuty	80	80	80	80	80	80	80	80	80	80	80	80	960
Jira	300	300	300	300	300	300	300	300	300	300	300	300	3,600
[itemized expense]													
Subtotal	1,330	1,330	1,330	1,330	1,330	1,330	1,330	1,330	1,330	1,330	1,330	1,330	15,960
Office & Misc													
Office supplies	100	100	100	100	100	100	100	100	100	100	100	100	1,200
Staff development	1,500	1,500	1,500	1,500	1,500	1,500	1,500	1,500	1,500	1,500	1,500	1,500	18,000
Conferences	500	500	500	500	500	500	500	500	500	500	500	500	6,000
[itemized expense]	80	80	80	80	80	80	80	80	80	80	80	80	960
[itemized expense]	300	300	300	300	300	300	300	300	300	300	300	300	3,600
[itemized expense]													
Subtotal	2,480	2,480	2,480	2,480	2,480	2,480	2,480	2,480	2,480	2,480	2,480	2,480	29,760
Total EXPENSES	98,810	92,810	92,810	97,810	97,810	97,810	97,810	102,810	102,810	107,810	107,810	107,810	1,204,720

Figure 7-1. Engineering budget template (https://oreil.ly/leX0A)

Insert your fixed and variable costs, including people, into the spreadsheet. The template breaks down costs monthly. This is typically how most folks track spending: short enough to be flexible, long enough to identify meaningful trends. You may or may not have to consider income. It's unlikely. Forecasting and measuring are generally finance functions that work with sales and marketing. It's included in our example template, but you can remove it if it's irrelevant.

We usually prepare three variant budgets for different scenarios. One for the best case, one for the expected outcome, and one for the worst case. This is contingency planning because, let's be honest, something always goes wrong (or occasionally goes amazingly well). That's why we need contingency planning. This isn't about being pessimistic; it's about being prepared. We need to consider specific risks, such as what happens if your largest customer leaves, and general uncertainties, like economic downturns. The key is having plans for different scenarios: best-case, worst-case, and "well, that was unexpected" scenarios.

Tracking and Managing

Creating a budget is just the first step; now we need to manage it effectively. We generally recommend reviewing your significant costs, particularly headcount, every month. Cloud services are an excellent example; tracking your spending monthly is a good cadence that should identify trends. Additionally, most cloud platforms have tools to set budgets and alert you to variances. We aim to conduct quarterly strategic reviews that examine the broader picture, typically with the entire organization. We usually produce a new budget annually.

When actual numbers don't match the plan (and they never do), we must understand why. This isn't about pointing fingers; it's about learning and adapting. We need to look for price variances that tell us if something cost more than expected, volume variances that show if we used more or less than planned, timing differences that indicate whether we spent money sooner or later than expected, and efficiency impacts that reveal whether we're getting more or less bang for our buck.

Once you have identified a positive or negative variance, share it. Hiding cost over-runs is never successful; it is much better to face that head on. In your organization, different people also need different levels of detail in financial communications. Your leadership will generally want a big-picture view that tells them managers need specific information about their departments to ensure we're on track. Team leaders require project-level details that enable them to make informed day-to-day decisions. The key is presenting information in a way that's useful for each audience without overwhelming them with irrelevant details.

We need to be flexible but not chaotic when handling budget changes. This means having transparent processes for minor adjustments, such as transferring funds between line items, making major changes in response to significant market shifts, and responding to unforeseen events in an emergency.

Learning from Experience

After each budget cycle, we need to learn from what happened. What worked well that we should keep doing? What failed that we should avoid next time? What surprised us that we should prepare for in the future? Most importantly, what did we learn that can help us do better next time? It's about continuously improving how we plan and manage our resources.

The goal isn't to predict the future perfectly; it's to have a plan that helps you make better decisions and adapt when things change. Because things will change, as they always do. The best budget enables you to achieve your goals while being flexible enough to adapt when reality doesn't match your plans. And reality never matches even the best plans, so we keep learning and adjusting as we go.

Vendor Management

Let's discuss something that might not sound exciting but can make or break your organization: vendor management. In today's world, we're all connected like a giant web of business relationships. Gone are the days when vendors were just people who sold you stuff; now, they're strategic partners who can either propel you forward or hold you back.

The Great Build-Versus-Buy Debate

Build-versus-buy is the art of identifying what to build as opposed to what to buy. We always focus on what we call "core business," which normally means building the things that provide a business edge or competitive advantage and not building the things that aren't core business. Here's a question that keeps many leaders up at night: should we build it ourselves or buy it from someone else? It's like the "make dinner or order takeout" decision, but with much bigger stakes. When making this decision, we must consider our core competencies alongside our available resources and time constraints. We also need to think hard about the long-term implications of our choice. Creating a custom solution might seem appealing, but we also create a future maintenance headache that will haunt us for years.

In Practice: April's Story

April faced this decision with PixelCurl's deployment pipeline. The StudioOps team had historically maintained a tailored customer improvement/customer development (CI/CD) system that was built by two senior engineers years ago. It consumes 20% of their time just for maintenance. April recognized that deployment pipelines weren't PixelCurl's competitive advantage: their innovative animation rendering engine was. She made the strategic call to migrate to a commercial CI/CD platform, freeing those senior engineers to mentor juniors on the actual core business.

Just because we can build something doesn't mean we should. Sometimes, the best technical solution isn't the best business solution. We need to be honest about our capabilities and limitations. Building in-house means committing to ongoing maintenance, updates, and support; it's not just about the initial development.

We always tell people to focus development efforts on core business, where building and owning something provides a business or competitive advantage. If you can buy a commodity service off the shelf, do it. Don't distract from your core business by building something that doesn't provide that advantage. This shouldn't be hard calculus either: creating your own database (unless you're a database company), your own scheduling system, or your own authentication and encryption are all unnecessary. It should be easy to determine that building these sorts of things is an indulgence, not a requirement.

Your vendors are the individuals who will provide those noncore capabilities, so they need to deliver the most effective, cost-efficient, and reliable services. A prime example of this in the technical world is cloud services and SaaS business tools. These have primarily replaced in-house datacenters, specialized software solutions have taken over where custom development once reigned, and external expertise fills the gaps that we can't (or shouldn't) fill ourselves. This dependency is both powerful and scary. With most vendors, you're not just looking for a quick transaction; you're potentially starting a long-term relationship that requires care, attention, and clear boundaries. Choosing the wrong partner can lead to painful experiences and force you to focus on noncore activities at the expense of your business.

Picking Your Partners

Selecting vendors is like hiring employees, except they come with their own companies attached. We need to look beyond just capabilities and track records. While those are crucial, we must also evaluate their stability and cultural fit. A vendor might have the perfect solution, but if their work style clashes with your culture, the relationship will be an uphill battle from day one.

You also need to work out your requirements (must-haves versus nice-to-haves), the longevity of the solution (and the vendor) and, importantly, what it costs. Contract negotiation isn't just about getting the lowest price; it's about creating a framework for a successful relationship. We need crystal-clear definitions of what we're getting, how we'll measure success, and what happens when things go wrong (because they inevitably will). The best contracts don't try to predict every possible problem but rather establish clear principles and processes for handling whatever comes up.

Managing the Relationship

The real work begins once we've picked our vendors and signed the contracts. Vendor management requires regular attention, care, and occasional pruning. Along with reviewing performance metrics, regular check-ins offer opportunities to share feedback, discuss upcoming changes, and catch minor issues before they become significant problems. A good vendor relationship should feel like a partnership, not a transaction.

Outsourced People and Teams

Outsourcing functions is like asking someone else to handle your organization's relationships; it's complicated and requires extra care and attention. Outsourcing peo-ple functions—whether near-sourcing, outsourcing domestically, or hiring offshore contractors—presents complex challenges and risks that require careful evaluation. When you entrust partners with traditionally managed in-house functions—such as human resources, recruitment, or employee relations—you relinquish control over critical processes that shape your culture and workforce. The partners may not fully understand the company's values, mission, or culture, which can result in both poor service delivery and erode morale and trust within your internal teams. A common problem is when a partner prioritizes skills fit, compliance, and transactional effi-ciency over finding individuals who align with the organization or team's culture and have the potential to grow and develop.

We always recommend aiming for relationships with partners and the engineers they provide, where those engineers feel like they are part of the team. This differs from an engagement with an agency, where a scoped project is involved. Those relationships have their place, but if you bring on people who will interact with your team, they should be treated like part of the team.

This is why we recommend never outsourcing your core hiring processes. Have the partner find people, but put them through your existing interview process. Don't just take the first resume off the pile and proceed. Trust, but *verify*, is the watchword when bringing in people from outside. This also reduces the potential impact on organizational culture. A company's (and a team's) culture is built on shared values, informal norms, and personal interactions, which can be disrupted when you bring in people who are perceived as "outsiders" or treated like components in a machine. The golden rule applies strongly here: treat external engineers as you'd like to be treated.

This also reduces the potential that your team may feel disconnected from decision-making processes or perceive the organization as prioritizing cost savings over their well-being. It reduces the risk that institutional knowledge will reside outside the team in the hands of external folks or that design decisions will be made without a broader organizational context, increasing the risk that you can no longer maintain core capabilities with your team if those people move on. Finally, it also reduces communication barriers, especially with offshore contractors operating in different time zones or cultural contexts.

This is never going to be perfect, but a lot of friction can be reduced by treating everyone equally. We both have one-on-ones and sometimes performance planning discussions with contractors, just like any other member of the team. We also try to balance time zones against collaboration. For example, in a recent role, James had teams located in Ukraine, so we held meetings in the morning Eastern Standard Time, which corresponded to the mid-to-late afternoon in Ukraine, ensuring overlap with the team there and enabling interaction between both groups. It's often hard to juggle this, but maximizing the time that the teams spend together improves collaboration, communication, and trust.

There are also legal, regulatory, and data privacy issues when outsourcing humans. Are they in a jurisdiction that has different privacy or data protection laws? Do you feel comfortable that their company will protect your data, especially if they use their own assets or the company has many clients? This refers to selecting vendors carefully: the security of your data and assets are baseline requirements for choosing a partner.

Outsourced people can offer efficiency and cost benefits, but there are also downsides. A good rule of thumb is that you get what you pay for, so weighing these advantages against the potential drawbacks to culture, employee experience, and compliance is essential. Thoughtful decisions about where you get your people from and how you manage them are key.

Ethics

Let's talk about ethics, because running a business isn't just about making money—it's about doing it the right way. We'll be blunt: doing things the right way matters because it's both the right thing to do and because not doing it right will cost you people, reputation and, ultimately, business. We frequently tell the people we're working with that if you don't want to do the right thing because it's the right thing, then at least do the right thing to avoid damaging your brand.

So, what do we mean by ethics? We group this into a few areas: taking care of people by treating people fairly and decently, acting honorably, and communicating transparently in our communities, both internally with our colleagues and externally in our user communities.

Taking Care of Our People

Our employees aren't just resources; they are people with lives, families, and dreams. Our budgeting and management decisions need to reflect that reality. This means ensuring fair compensation that enables people to live with dignity, investing in their development because they need opportunities to grow, and maintaining safe working conditions. Nobody should get hurt making a living.

In Practice: April's Story

April discovered a pay gap in the StudioOps team: the junior engineers, despite their struggles without mentorship, were being paid 30% below market rate with the justification that they're "learning on the job." She argued successfully that proper mentorship and fair compensation aren't mutually exclusive; in fact, underpaying while under-supporting is doubly unethical. Her revised budget included both market-rate adjustments and dedicated mentorship time.

Being a Good Neighbor

Our organizations don't exist in a vacuum. Every decision we make has ripple effects on our communities and environment. Since we have only one planet, we must consider our environmental impact, reflect on how our actions affect our local communities, and take responsibility for giving back in meaningful ways.

The Global Perspective

In our interconnected world, we must navigate different cultural expectations with care and respect. What's expected in one place might be inappropriate in another, but we still need to maintain consistent ethical standards across our operations. This means respecting local customs while ensuring fair labor practices and ethical behavior everywhere we operate.

Looking to the Future

Ethics aren't static—new challenges emerge constantly. The rise of artificial intelligence and automation raises new questions about intellectual property, responsibility, and accountability. Data privacy concerns grow more complex as technology advances. Sustainability becomes increasingly crucial as we face global environmental challenges. We need to stay ahead of these issues, thinking beyond the next quarter to consider the long-term implications of our decisions.

Making It Real

Ethical principles are great, but they must be more than words on a wall. We need to regularly train our people in ethical decision making and create an environment where people feel safe speaking up when something's wrong. James worked at Microsoft, where a comprehensive program explained exactly what Microsoft expects from employees and provided articulated scenarios to demonstrate these expectations. Everyone in the organization was fully aware of their obligations and knew what to do if they observed unethical behavior. Many companies provide this training, and you can select a vendor with an appropriate program. This means learning from our mistakes and adapting our approaches as new challenges arise.

Remember: ethical behavior isn't just a nice-to-have but is essential for long-term success. In today's connected world, our organizations are only as strong as our weakest relationship and our smallest ethical lapse. While we can't be perfect, we can strive for continuous improvement and conscious decision making. When in doubt, we should ask ourselves: would we be proud to have this decision on the front page of tomorrow's newspaper? If the answer is no, it's probably time to rethink things.

Conclusion: Budget as Strategy

In this chapter, we've navigated the often-unfamiliar territory of budgeting, costs, and vendor management for engineering leaders. We've seen that budgeting isn't just a financial exercise but a strategic tool that helps align your resources with your organization's goals. The bridge connects your grand visions to practical reality, enabling you to anticipate challenges, seize opportunities, and make informed decisions about your team's future.

We explored the nuances of cost management, understanding the difference between fixed costs (your organizational foundation) and variable costs (your flexibility factors), and how step-up costs can surprise you when you level up. We discussed the importance of tracking who pays for what and building a culture that considers costs as part of the decision-making process.

The budget creation process might seem daunting, but starting with historical data, categorizing expenses thoughtfully, setting sensible goals, and having a flexible template can make it manageable. Remember, the goal isn't perfection; it's having a living document that guides decisions and adapts as circumstances change.

Vendor management is also critical, with the ever-present build-versus-buy debate at its center. We emphasized focusing your development efforts on the core business, where you gain a competitive advantage while partnering with vendors for everything else. When outsourcing people and teams, we highlighted the importance of treating external engineers as part of the team to preserve culture and knowledge.

Finally, we touched on ethics—doing the right thing, not just because it's right, but because it makes good business sense. Taking care of our people, being good neighbors, maintaining consistent ethical standards globally, and anticipating future challenges are all part of creating an organization that stands the test of time.

Budgeting, costs, and vendor management might not have drawn you to engineering leadership, but excelling in these aspects will give you the foundation to deliver on the technical vision that did. By bringing the same rigor and thoughtfulness to these business problems that you bring to technical ones, you create the conditions for your team and your organization to thrive.

It's also a skill you'll use more and more as you grow your career, both as a leader and as a more senior individual contributor. When we first started in the industry, neither of us ever imagined spending the amount of time we now spend caring about budgets and vendors. We both now lead technical and product teams, but these skills have been crucial even when we've dipped back into being senior individual contributors. To build good software, you need a holistic view of how that software is built, including what it costs.

In Chapter 8, we'll introduce you to technical strategy and how to use it to drive execution and direction.

Technical Principles and Strategy

The phrase "developing and implementing technical principles and strategies" sounds fancy—like the title of a 200-slide deck that someone with "enterprise architect" in their job title puts together. In reality, it's just a fancy way of saying, "What do our technology, principles, architecture, and capabilities need to look like, and how do we get them there to meet our business goals?"

More formally, technical strategy is the comprehensive plan that aligns technology decisions, architecture choices, and capability development with business objectives, providing a framework for making consistent technical decisions that support organizational goals.

Technical principles and strategy (from here on, we'll just say "strategy") are part of a roadmap—and indeed, they should be included in your current roadmap. The product features that you plan to build must be supported by the technology and architecture underpinning them. If you're planning a significant new feature that needs more scale, then your technical strategy must address how that scale will be achieved. It's also a blueprint—though in many chaotic organizations, it's one that's frequently updated and revised. Part architecture document, part register of tools and their selection criteria and evolution, it provides the foundation for planning the addition of people and specific skill sets to your team.

In this chapter, we explore the core principles, frameworks, and practices that underpin technical strategies. We'll give you a practical look at how to use technical strategy to tackle pesky challenges, like thinking about what's next and how to get there, dealing with legacy systems nobody wants to touch, or ensuring you're constantly evaluating what works and what doesn't. This is a practical guide for teams living in worlds where the only thing you can count on is that everything will change tomorrow.

Why Does Technical Strategy Matter?

Let's face it: a well-thought-out technical strategy is crucial if you want your organization to grow without imploding. Your technical strategy ensures that all those technical and engineering decisions—from big-picture architecture stuff down to "Which JavaScript framework should we use this week?"—align with what your company is trying to accomplish. This alignment isn't just nice to have; it ensures that every tech investment and process tweak is helping you meet your business goals. In other words, you develop your technical strategy by weaving it into your organization's game plan, connecting your tech choices with what you're trying to achieve in the market.

In today's world and in most, if not all, chaotic environments, any rigid plan will be obsolete before you finish your morning coffee. Technical strategies need to be flexible and iterative. With a flexible approach, you can move people and shift priorities without significantly disrupting ongoing projects. There's always a balance that needs to be struck between being able to move quickly and avoiding flailing.

Remember that strategy is more than what widgets we should buy or build; nonfunctional needs, like security, compliance, and availability, must be addressed by your strategy. For a lot of chaotic organizations, we see two modes for nonfunctional needs: we ignore them or we overcompensate for them.

A technical strategy means asking yourself important questions. For example, what do you need for security, compliance, or availability? Review whether you have customers in the European Union, meaning General Data Protection Regulation (GDPR) must be considered. Or do most of your customers expect you to have Systems and Organization Controls 2 (SOC2) compliance standards in place? Do you need two cloud regions or enhanced availability because you've promised customers a specific uptime? Perhaps those seem like reasonable investments that address real needs, which you can probably afford without breaking the budget and constraining your ability to iterate and ship software. Your risks and constraints may differ in larger organizations, and your strategy should reflect that. A solid technical strategy is a good place to document your thoughts about today's needs and avoid premature optimization.

Another important piece is ensuring your technical and product strategies play nicely together. In early-stage companies, these are often inseparable; your technical choices directly enable or constrain your product capabilities. As organizations mature, product and technical strategies may diverge but must remain aligned. While product strategy focuses on what needs to be built and why you're building it, technical strategy provides the framework for how you will build those products. You'll need both to build products that meet business goals even when everything around you changes at warp speed.

Building a Foundation

You must know where you stand before you can aim for the stars. Understanding your organization's current landscape is an essential starting point for any technical strategy. That journey starts with putting together an inventory. It'd be wonderful if this inventory contained detailed systems documentation, equipment inventories, network configuration diagrams, and the policies and procedures governing their use. (We know; we think that's funny too.) It's far more likely your journey will start with a list of applications, languages, frameworks, and tools. Your mission is to evolve that inventory into (more) mature documentation. Indeed, far from being some static record gathering dust on a shelf, this inventory is the baseline for every architectural decision and technology investment.

Based on his initial conversations with the team, Figure 8-1 is the initial systems architecture diagram that James created on his first day at a startup in the past.

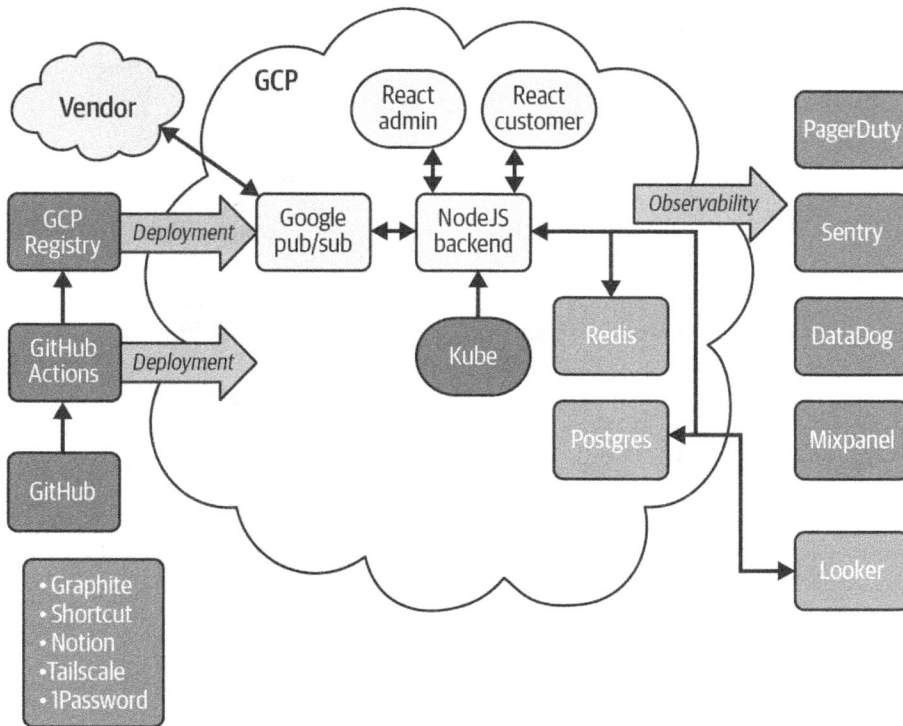

Figure 8-1. An example of initial systems architecture

It's not rocket science, overly time consuming, or complicated. It's an attempt to outline the architecture, some of the workflow, and key tools that are essential to the environment. By the end of the first week, James had done the following:

- Drawn all over this diagram to add notes, and added more workflows
- Documented all (most) of the other tools being used and identified any tooling gaps
- Identified single points of failure
- Matched systems and tools to internal owners and experts (e.g., Rae as the expert on the backend, Jeff as the subject matter expert on the React frontend, and Jess managing the CI/CD workflow)
- Started to discuss the strengths and weaknesses of specific subsystems (e.g., less than 50% test coverage for the backend)
- Worked out if there were any documentations, procedures, or policies for the operation of the environment (spoiler: there was not a great deal)
- Worked out where they stood on nonfunctional requirements like security and compliance
- Started organizing this information into a structure, identifying gaps in his knowledge, and formulating a strategy for what to tackle first

With a clear(-ish) picture of your existing systems, you can assess capabilities, identify gaps, and prioritize what needs to be done. For example, understanding your current application, network architecture, and system dependencies helps pinpoint bottlenecks or single points of failure that might make scalability a nightmare. Similarly, having some idea of how things work—even if it's just rough flowcharts and diagrams rather than well-documented policies and procedures—ensures that changes or upgrades are more likely to be smooth and, sometimes, more likely to be feasible. This reduces the risk of things going wrong and paves the way for smooth improvements.

You Need More than Technical Widgets

The technical inventory is just one piece of the puzzle. Leaders also need to understand the human and process-related dimensions of their current systems. This need means understanding team expertise, current workflows, and how various people, processes, and components depend on each other. Ultimately, technical strategies are implemented, and the results are operated by the humans who work with you. You need to understand how things work now and who does them, and then ensure that whatever you introduce in your technical strategy can be implemented and operated. The most brilliant technical strategy in the world that your team can't execute is just a nice but useless artifact.

Some years ago, James created Figure 8-2, a mind map of the areas he asked about when he started at a new organization. Some of them are more relevant than others, depending on your role and the scale of your organization. They aren't a definitive list of areas to explore, but they provide a good starting point.

While the mind map James created covers four major domains (company, product, technical, and humans), the deep dive in this book on technical strategy focuses primarily on the technical domain. Earlier chapters cover the organizational, product, and human aspects in detail. Here, we'll assume you've done that groundwork and can now focus on the technical strategy itself.

You can view the mind map online (*https://oreil.ly/supp-eng-lead*) for a closer look and adapt the Xmind and SVG copy of this for your purposes.

CTO / VP of Engineering: the first 90 Days

Company
- What are we committing to?
- Our metrics
- Company metrics
- Mission
- Budget/Runway

Product
- Planning
  - Current Roadmap
  - Prioritization
- Inputs
  - Support
  - Leadership
  - Customers
  - Partners
  - Engineering
- Outputs
  - Org
  - External
    - Customers
    - Partners
- Tracking
  - Data
  - Reporting
  - Accuracy
- Process
- Developer Experience
  - Cycle time
  - Code review
  - Push
  - Feature flags
  - Metrics
  - QA

Humans
- Recruiting
  - Evangelism
  - Interviewing
  - Onboarding
  - Pipeline
- Performance
  - Morale
- Mentoring
  - Education
- Capacity
  - Forecast
  - Retention
- Structure & Org
  - Ladder
  - Roles
  - Seniority Distribution
  - Structure
- Communication
  - 1:1s
  - Team meetings
  - Product/Eng/Design Comms
  - Org Comms
  - External
  - Code Review
- Costs
  - Salary
  - Stock
  - Benefits
- Diversity & Inclusion
  - Recruiting
  - Levelling

Technical
- Ops
  - Monitoring
  - Security
  - DRP
- Platform
  - Architecture
  - API
  - Payments
- Native
  - Android
  - iOS
- Mobile
  - PWA
- Infrastructure
  - Costs
  - Hosting
- Deployment
- Testing
  - Feature & AB Testing
  - Infrastructure
  - Application
  - Quality
- Capacity
  - Traffic
    - Breakdown
    - Behavior
  - Performance & Scale
- IT
  - Onboarding
  - Internal IT
- Support
  - On-call
  - Bugs & Backlog
- Knowledge Management
  - Communication
  - Documentation
  - Events
- Data
  - Collection
    - Schema
  - Storage
  - Visualization
- Incidents
  - Process
  - Post-mortems
- Security
  - Compliance/Privacy
  - AAA

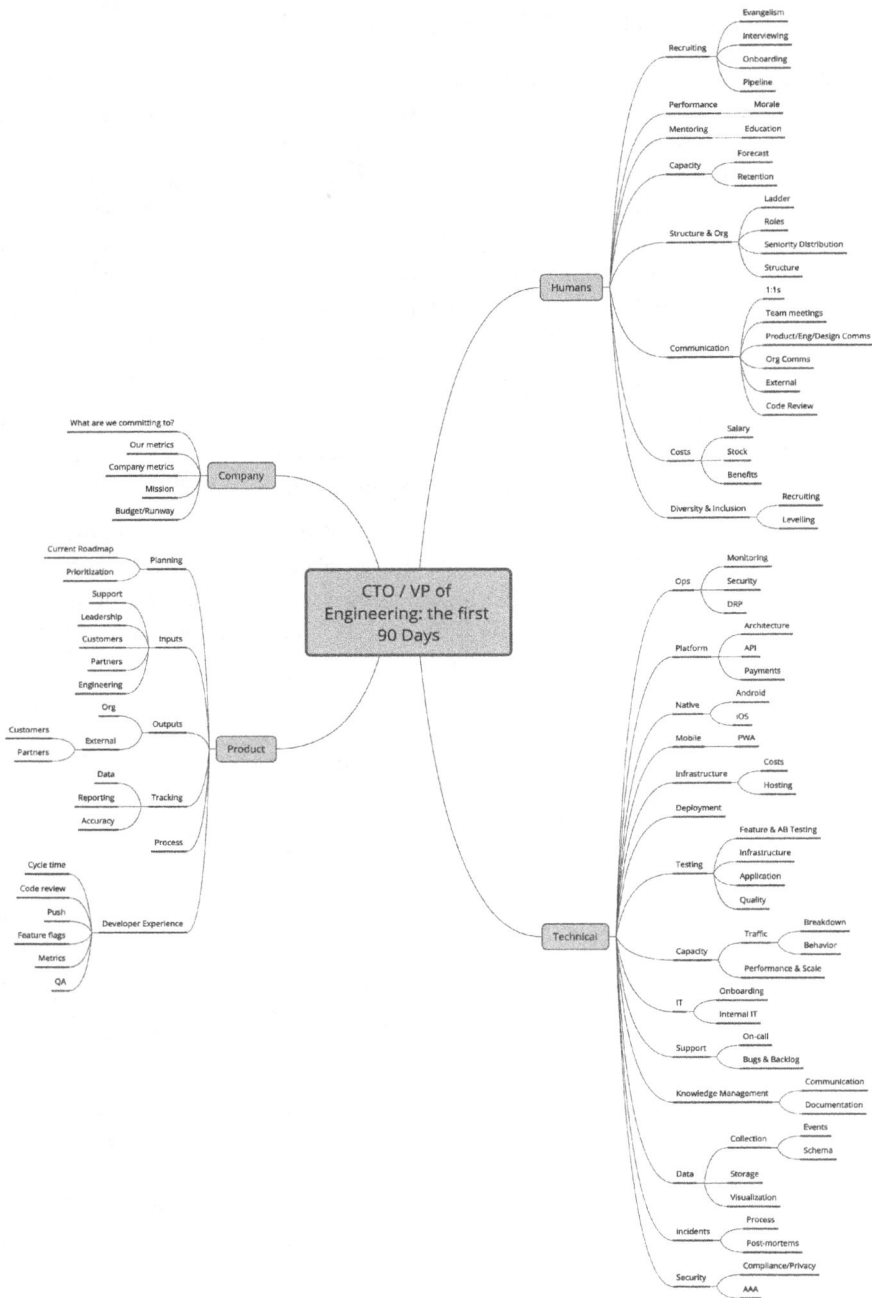

Presented with xmind

Figure 8-2. Mind map of a chief technical officer and vice president of engineer's first 90 days in a new organization

Not every item on the map is relevant to every organization, at least not at every stage of evolution. Instead, it serves as a helpful mental checklist. We recommend skimming through most of the items and reaching one of three conclusions in relation to your organization:

1. No action is needed; everything is going smoothly.
2. This needs work; let's determine what to do and when.
3. There's nothing here, and should that worry me?

You can then determine what's broken, prioritize what needs to be fixed, and watch for emerging issues. You then combine this understanding with your technical insights into a foundational view of how things look now. With this, assuming you don't resign in horror, you can work toward a strategy for tackling the key items.

Establishing Technical Principles

At the heart of every effective technical strategy are technical principles. We use these principles to guide all our strategy's engineering decisions. The best technical principles are clear, concise, and easy for everyone in the organization to understand—whether you're a seasoned engineer or the new hire who started last Monday. They ditch complex jargon in favor of language that makes sense to both technical and nontechnical stakeholders.

A key characteristic of practical technical principles is that they're forward looking and aspirational. They must be robust enough to support long-term growth while still adapting to your organization's evolving needs. In practical terms, while the principles must be specific enough to guide everyday decision making, they must also be flexible enough to accommodate new ideas. We've discussed how chasing new shiny things can be dangerous and counterproductive, but sometimes change makes sense. If your technical principles restrict thinking about innovation and evolution, your team will bypass or ignore them. For instance, principles that emphasize modularity and loose coupling in software design not only streamline current development efforts but also make it way easier to integrate new technologies as they become available.

Your technical principles are also likely to be aspirational. Some principles may be reflected in you, your team, and your environment. Alternatively, you may have parts of your organization where what you've built aligns with the ideal expressed by your principles. But much of your environment is often a cautionary tale of what happens when you don't follow technical principles. So, does this mean technical principles are empty words? Absolutely not—technical principles have a powerful two-fold function in this instance.

First, they stop you from making things worse. It's easy to throw good money after bad, but if you use your technical principles as a compass, then they will help you avoid making a worse decision on top of a bad situation. They give you an aspirational target and define what good looks like, so you know where you need to aim when developing your strategy and when implementing that strategy.

Second, the (massive) bonus prize you also get from technical principles is that they don't just guide your strategy; they are reference points that help you make good decisions throughout your product and development lifecycle. They live beyond your strategy, with the best technical principles being the ones that you reference when you review a pull request, read a product design, or examine a new technology. You can also go back into creating your technical principles by bubbling up the principles that your team thinks make good software and culture. We go into this in more depth in Chapter 9, where technical principles also help us make good changes.

Example Technical Principles

So, what technical principles should you have? It's hard to create a comprehensive list for every organization at every stage of evolution. But this section outlines an example list of 14 practical principles that a software engineering team might want to adopt.

These principles represent both aspirational goals and practical trade-offs. Early-stage teams might adopt fewer principles with more flexibility, while mature organizations might enforce stricter adherence. Each principle directly supports business goals by either reducing costs (through maintainability and automation), increasing customer satisfaction (through reliability and performance), or enabling growth (through scalability and modularity):

Simplicity first
Prefer straightforward, clear solutions over complex ones. This reduces support costs and accelerates onboarding.

Scalability
Design systems that can scale horizontally and vertically. This enables customer growth without architecture rewrites.

Modularity
Build your systems in independent, interchangeable components. This enables parallel development and reduces time to market.

Reliability
Prioritize fault tolerance and resilience in your designs. This increases customer retention and reduces churn.

Maintainability

Write code that is easy to understand, modify, and debug, and ensure consistency through standards, automation, and validation. This will reduce long-term costs.

Documentation

Treat documentation as integral, not optional. This accelerates onboarding and reduces knowledge silos.

Testing

No untested code will be merged and no code that fails tests will be merged. This rule reduces defects and improves customer satisfaction.

Visibility

Observability and metrics are core requirements, not afterthoughts. This principle enables data-driven decisions.

Security by design

Incorporate security considerations at every stage. To reduce breach risks and maintain customer trust, security must not be an afterthought.

Performance awareness

Balance optimization efforts with practical requirements to improve user experience.

Consistency

Use established patterns and standards consistently. This reduces cognitive load and speeds up development.

Automation

Automate repetitive tasks whenever possible to reduce operational costs.

Openness

Design for interoperability and openness, using open standards. This avoids vendor lock-in.

Evolution

Architect systems anticipating change and evolution. This reduces future refactoring costs.

Many of these principles will resonate with you and your team. Indeed, we recommend using these examples as a starting point for discussing technical principles. But building your technical principles should be a collaborative effort—created by and living inside your team.

Crafting Technical Principles Workshop

Let's walk through a practical workshop format that we've found effective for teams developing their technical principles. This structured approach helps ensure that your principles emerge from collective wisdom rather than individual preference.

Exercise: Crafting Technical Principles for Startups

Objective

The goal is to collaborate and develop technical principles that will guide a startup engineering team in making effective, aligned decisions as the startup grows.

Materials Needed

- An online collaborative whiteboard tool (such as Miro, Mural, or FigJam)
- A timer to keep the workshop focused and energetic (Note that the suggested times are just that—suggested. If you have a larger or smaller team, you can adjust them.)

Preparation

Before starting, ensure everyone understands why technical principles matter (i.e., they serve as decision-making guardrails that align the team and provide consistency over time).

Exercise Steps

Step 1: Idea dump (10 minutes)
> Have each team member use the online collaborative board to post their ideas for principles as individual sticky notes. Encourage everyone to think broadly. What values should guide technical decisions? What have they learned from past experiences? This is a judgment-free brainstorming phase, so quantity matters more than quality at this stage.

Step 2: Grouping and naming (10 minutes)
> As a group, start categorizing the posted principles into themes. Each theme could then be summed up with an overarching principle. For example, sticky notes like "Prioritize scalability" and "Optimize for performance" might fall under a broader category, such as "Build for growth." This consolidation helps identify patterns in the team's thinking and reduces overlapping concepts.

Step 3: Voting (5 minutes)
> Give each participant three votes to place on the principles or themes they find most crucial. Votes can be distributed however individuals choose—all on one principle or spread across multiple. This democratic approach ensures the final principles reflect the collective priorities of the team.

Step 4: Drafting principles (15 minutes)

Break into smaller groups or stay as one team (depending on the size of your group) to draft well-articulated versions of the top-ranked principles. For each principle, define what it means, why it's essential, and provide a concrete example to make it tangible. The goal is to transform general ideas into actionable guidance that can inform day-to-day decisions.

Step 5: Discussion and ranking (15 minutes)

Discuss the top-voted principles as a team, allowing people to advocate for principles they believe are essential. Rank the principles based on importance and relevance to your specific organization's needs and challenges. This prioritization is beneficial when principles come into conflict during decision making.

Step 6: Review (5 minutes)

Examine the finalized list of principles, ensuring they are clear and actionable, and that they genuinely reflect the team's values. Create a document outlining your principles and share it widely. Consider integrating them into your architecture decision records, code review checklists, and onboarding materials to ensure they're referenced and utilized.

This workshop approach transforms the abstract concept of technical principles into a concrete set of guidelines with broad team buy-in. The collaborative nature ensures that diverse perspectives are incorporated and that the final principles reflect the collective wisdom of your engineering team.

In Practice: April's Story

In the StudioOps team, April ran the technical principles workshop with a twist. She started by having the team list their current pain points: Marcus complained about debugging Sarah's frontend code without documentation, Sarah couldn't understand Derek's database-stored procedures, and Derek was frustrated by Marcus's authentication system that required manual restarts. From these complaints, they established principles: "documentation before deployment" (no pull request merges without README updates), "boring technology first" (addressing their React version chaos), and "observable by default" (as they currently have no monitoring on half of their services). The senior engineers initially resisted the principles until April showed them that they had spent hours that quarter just figuring out each other's code patterns. That's potentially months of an engineer's time lost to inconsistency.

Living Your Principles

To ensure these principles are lived, they also need active endorsement from leadership and champions across the team. This is often where those pesky staff and principal engineers earn their money, championing technical principles. However, it's essential that all engineering leaders—from chief technology officers to engineering managers, as well as staff and principal engineers—must champion these technical principles. They ensure they become embedded in the organizational culture and influence every decision, from high-level strategic planning down to the nitty-gritty details of code development. This top-down endorsement helps ensure that the principles aren't just dusty documents but are consistently followed.

Scaling Without Overengineering

After you have your foundation and principles, the most significant objective for your technical strategy is to let you scale without overengineering. Scaling without going overboard on engineering starts with a straightforward principle: keep it simple! When designing your strategy and architecture, start with something simple. The whole approach comes down to the idea that simplicity is much more sophisticated than complexity. By avoiding unnecessary complications, you reduce the risk of introducing headaches and problems later on. You need to think about future scalability, but your strategy should mainly tackle today's requirements. Over-optimizing for some hypothetical future that might never happen can waste resources and make things way more complex than they need to be. Instead, let your scaling decisions be guided by real-world demand and data, ensuring that any changes respond to real needs rather than hypothetical scenarios.

This isn't to say you shouldn't think about future needs, but it's important to strike the right balance. Investing in good technology means planning for what you need now and where you might expand later. You likely don't need to build the most advanced or complex world right off the bat; what's essential is having a clear plan that lets your system evolve gracefully over time. This approach helps ensure that your system can scale efficiently as your business grows and that you can make incremental improvements without breaking everything.

The key trade-off decisions in scaling come down to differentiation versus commodity and just-in-time scalability. Choose to build only what differentiates your business; buy everything else. Scale only when you have evidence that you need to, not based on optimistic projections.

An excellent example is database design and selection. If you're building most applications, choosing PostgreSQL as your first database is a strong choice. It's performant, scales well, is well represented and supported by cloud providers, and has a healthy ecosystem for ORMs (object–relational mapping), tooling, and support for numerous languages and frameworks. There is extensive documentation and many examples online that provide guidance on how to work with, tune, and configure it for various scenarios. It's almost "you won't get fired for buying IBM" good. Combine this with a well-designed schema—for example, typing integer fields that might grow as large as `bigint` instead of `integer`[1]—and you can achieve a solid solution without unnecessary complexity, investment, or overhead. You're probably pretty well set up if your product takes off.

On the other hand, different choices make sense for different contexts. SQLite excels for embedded applications or development environments. In addition, CockroachDB becomes valuable when you genuinely need global distribution and can afford the operational complexity. The key is matching the technology to your actual requirements, not your imagined future state.

In Practice: April's Story

April discovered that the StudioOps team was running MongoDB for user data, PostgreSQL for application data, Redis for caching, Elasticsearch for logs, and DynamoDB for session data. When she asked why, the answer was always "That's what [specific senior engineer] knows best." The monthly Amazon Web Services (AWS) bill was $45,000, with $12,000 just for running five different database systems. April's strategy was to consolidate to PostgreSQL for everything except caching (keeping Redis). PostgreSQL can handle JSON documents (replacing MongoDB), has full-text search (replacing basic Elasticsearch use cases) and, with proper indexing, can easily handle session storage (replacing DynamoDB). This isn't about PostgreSQL being perfect; it's about reducing operational complexity. The team can now gain a deep understanding of one database, rather than trying to understand five. The AWS bill dropped to $31,000, and deployment time improved by 40% as they weren't coordinating across five different datastores.

In addition to keeping things simple, leveraging SaaS, off-the-shelf, and cloud-based solutions is a game changer for building scalable systems without overengineering. Modern cloud platforms like AWS, Google Cloud, and Microsoft Azure provide elastic and flexible infrastructure options that let you roll with unexpected demand spikes. These services enable you to scale resources up or down as needed, so you don't have to build some crazy complex architecture that tries to predict every possible future scenario. This approach saves resources and makes your system more adaptable to real-time usage patterns.

---

1 The `bigint` type has a larger range than the `int` type (−2147483648 to +2147483647 compared with −9223372036854775808 to +9223372036854775807).

Making the right build-versus-buy decisions is also key to a good strategy. Throughout the book, we've discussed buying products and services that are not your core business. If you're not in the business of building a database, don't build a database. This seems self-evident in cloud services like computing and storage, as well as adjacent services like workflows, ticketing, knowledge management, and email, but don't stop there. Juan works in a large corporation with a strong focus on customer service. That organization developed tools to automate customer management tasks, including ticketing and service management. These functions are bread and butter for most service organizations and don't represent business differentiation. A new strategy—buying the right tools instead of building them, and focusing on the quality and performance of service delivery—was introduced to correct that.

The term "adaptable architecture" means systems designed to accommodate change without requiring complete rewrites. This includes using microservices where appropriate (but not by default), implementing proper abstraction layers, maintaining clear API boundaries, and choosing technologies with strong ecosystems that can grow with you. The goal is to build systems that can evolve incrementally rather than requiring periodic massive overhauls.

Choose Boring Technology

Every technical decision involves trade-offs. If you want more scale, that might cost more. If you want better performance, you might have to sacrifice some flexibility. If you want to use the latest tech, then you'll have a smaller hiring pool of engineers who know it. Be explicit about what you're prioritizing—and why—throughout your strategy development and implementation.

A particularly important lesson for engineers, especially those making technology choices for the first time, is to resist the temptation of chasing shiny new tech just because it's trendy. There's a common pitfall where a team feels they need to rewrite their entire codebase in the hottest new programming language or adopt the framework everyone is talking about on Hacker News without considering the long-term benefits. The engineer Dan McKinley has a maxim: choose boring technology (*https://oreil.ly/Ok-Jv*). He posits that you have a certain number of "innovation tokens." You spend an innovation token each time you make a technology choice that isn't "boring":

> Let's say every company gets about three innovation tokens. You can spend these however you want, but the supply is fixed for a long while. You might get a few more after you achieve a certain level of stability and maturity, but the general tendency is to overestimate the contents of your wallet. Clearly this model is approximate, but I think it helps.

If you choose to write your website in NodeJS, you just spent one of your innovation tokens. If you choose to use MongoDB, you just spent one of your innovation tokens. If you choose to use service discovery tech that's existed for a year or less, you just spent one of your innovation tokens. If you choose to write your own database, oh god, you're in trouble.

Any of those choices might be sensible if you're a javascript consultancy, or a database company. But you're probably not. You're probably working for a company that is at least ostensibly rethinking global commerce or reinventing payments on the web or pursuing some other suitably epic mission. In that context, devoting any of your limited attention to innovating ssh is an excellent way to fail. Or at best, delay success.

What counts as boring? That's a little tricky. "Boring" should not be conflated with "bad." There is technology out there that is both boring and bad. You should not use any of that. But there are many choices of technology that are boring and good, or at least good enough. MySQL is boring. Postgres is boring. PHP is boring. Python is boring. Memcached is boring. Squid is boring. Cron is boring.

The definition of boring technology naturally evolves. These days, we'd probably not say Node.js was a risky choice (though your mileage may vary), but the underlying framework is still consistent. In modern times, we'd put choices like "We should use Kubernetes" or "Gee, we need a service mesh" into that innovation token bucket; again, your mileage may vary.

You should resist the superficial appeal of jumping into tech that promises (potentially tiny) gains in innovation but often results in significant overhead, instability, and technical debt. Constantly rewriting code to keep up with the latest trends can lead to a ton of unnecessary work, and it might undermine the stability of your product and deeply frustrate your team. You've got to carefully weigh the potential benefits against the costs, ensuring that any move to new technologies is justified by improved performance, easier maintenance, or better scalability.

Every effort you invest into refactoring is also an effort that you are not investing in your product(s) and their development. We discussed the whys of prioritization in Chapter 6; this is a good time to ensure that you factor those into any architectural decisions you make.

Ultimately, your programming language or framework choice should be driven by your business needs rather than what looks cool in a blog post. The most modern or hyped technology isn't always the best fit for every organization, especially considering its impact on team dynamics, hiring, and long-term maintainability. A smart strategy is to optimize and refine what you have already when it makes sense, rather than embarking on a complete rewrite that might not yield enough advantages to justify all the blood, sweat, and tears. Keeping business objectives front and center ensures that your technology choices align with your strategic goals and provide a stable foundation for sustainable growth.

In Practice: April's Story

April faced this directly when Marcus proposed rewriting the authentication system in Rust "for better performance." April did the math: their auth service handles 1,000 requests per second at peak, with p99 latency of 50 ms using boring old Node.js. Moving to Rust might drop that to 20 ms, but at what cost? They'd need to hire Rust developers (currently zero on the team), maintain a separate deployment pipeline, and train the juniors on yet another language. April framed it in innovation tokens: "We're already spending one token on our custom animation rendering engine—that's our competitive advantage—and another on the real-time collaboration features that differentiate us. Do we really want to spend our last token on making authentication 30 ms faster when customers have never complained about auth speed?" The proposal was withdrawn, and they instead spent a week optimizing the Node.js implementation, achieving 35 ms p99 with technology the whole team understands.

Adopting new technologies should create specific, measurable value: reducing operational costs, increasing development velocity, expanding addressable markets, improving customer satisfaction metrics, or enabling new revenue streams. If you can't articulate how a technology choice drives one of these outcomes, you're probably spending innovation tokens unwisely.

Scaling without overengineering involves striking a balance, valuing simplicity, leveraging scalable cloud services, and (carefully and cautiously) investing in adaptable architecture. It requires a disciplined look at whether adopting new technologies (spending those innovation tokens) adds value or introduces unnecessary complexity. Keeping business needs in focus ensures that the tools and frameworks you choose enhance operational efficiency and long-term stability rather than being pursued because they're the latest fad.

Developing Your Technical Strategy

We've got the pieces of our strategy now: insights into our current world, a model for building without overengineering, and a set of technical principles. From these, creating a technical strategy is like planning a holiday; you need to know your destination, understand how you get there, and prepare for unexpected detours. At its core, a solid technical strategy connects your technology decisions to your company's big-picture goals, ensuring that everything you build supports the broader vision. When done well, it aligns your product vision (what you want to build) with your technical approach (how you'll build it), keeping your team focused and your products on target.

We like a well-planned trip. Indeed, James' travel planning spreadsheets are famous among our friends, primarily for their multilayered contingency management. We know, however, that you can't plan for every contingency in chaotic environments. So, in a chaotic environment, a strong technical strategy isn't static; it's dynamic, fluid, and adapts to changing conditions. This manifests most clearly in your choice of planning time frames, windows, and granularity.

Choosing Granularity

We recommend a practical time window. For many chaotic organizations, a quarter is a lifetime, and for others, planning for an entire year is feasible (while we question that, it's possible). You need enough time to invest in developing a meaningful and aspirational strategy, but not so much that the objectives become less aspirational than delusional. If you do want to look at the future, then dial down the level of granularity—for example, with detailed granularity in Q1, medium granularity in Q2 and Q3, and low granularity in Q4. Review the plan monthly to ensure you're on track and that the strategy aligns with reality.

In Practice: April's Story

April's StudioOps quarterly planning demonstrates practical granularity:

Q1 (high detail)
Consolidate to PostgreSQL (weeks 1–4), implement monitoring on all services (weeks 5–8), and standardize API patterns to REST (weeks 9–12).

Q2 (medium detail)
Complete authentication system consolidation, begin frontend standardization to React 18, and improve the CI/CD pipeline.

Q3 (low detail)
Scale infrastructure for expected customer growth, and address remaining technical debt.

Q4 (directional only)
Establish platform stability and optimization focus.

She reviewed this monthly, and by week 6 realized the PostgreSQL consolidation was taking longer than expected. Rather than rushing, she adjusted Q1 to focus on just MongoDB-to-PostgreSQL migration, pushing the DynamoDB work to Q2. In this example, you can see how the strategy remains intact, but the tactics adapt to reality.

Balancing Flexibility with Direction

Remember our discussion about scaling without overengineering? The same principle applies to your technical strategy. You want just enough structure to provide clear direction without becoming so rigid that you can't adapt when surprises pop up (and they always do).

Think of your strategy as a compass rather than a GPS with turn-by-turn directions. It points you in the right direction while leaving room to navigate around obstacles as they appear. This is especially important when we reflect on our technical principles; they serve as guardrails that help you stay on track, not concrete barriers that lock you into a single path.

Collaborative Development

Remember that the best strategies aren't crafted in isolation. As we recommended a collaborative approach to developing technical principles, your strategy should be designed with input from your team. The engineers who will implement the strategy, the product folks who will build on top of it, and the leadership who will support it should all have a voice.

This doesn't mean making decisions by committee (because that's a recipe for the most watered-down, least useful strategy possible). But it does mean getting diverse perspectives and buy-in before you finalize your approach. The people implementing the strategy need to understand and believe in it; otherwise, it will just gather digital dust in your shared drive. We explore this more in Chapter 9, where we go into depth into collaborative decision making.

Elements of a Technical Strategy Document

In this subsection, we'll walk through what a solid technical strategy and architecture document should include, with brief explanations of each section.

Introduction and overview

Clearly state the document's purpose, audience, and intended outcomes. Provide context so your audience, both within your team and external to it, can immediately grasp its significance and relevance.

Strategic vision and objectives

Articulate how technology directly supports the organization's broader mission and strategic priorities. Linking technology decisions to strategic goals clarifies the rationale behind every technical initiative. Suppose your organization has goals, such as quarterly sales goals. In that case, this is where you indicate how your work contributes to that objective or falls under product initiatives that support those goals. For example, we can't release version 2 of the product unless we can support more users. If you don't have strategic goals in the larger organization, then you might tie your goals directly to your product roadmap.

Current technology landscape

Document existing technologies, tools, and systems, including their current strengths and weaknesses. This is where the foundational insights you gain are placed, and as you improve the environment, this section is refreshed with the current reality.

Future state landscape

This is the core of your strategy; the list of things you will do. Here, we describe your vision for the future state landscape: what tools and technologies will be deprecated, what will evolve, and what will be decommissioned. We also like to include future state operations and processes—for example, if you don't have an incident management and incident review process, this is where you'd specify that you intend to have one. Like the previous section, this section evolves as your strategy is implemented.

Technical principles

This is where your technical principles live or are cited. Refer back to the principles we discussed earlier, the ones that guide your engineering decisions and help you avoid making things worse, even in chaotic environments.

Implementation roadmap

Provide a phased plan outlining how to transition from your current to your desired state. Chapter 6 discussed prioritization and planning extensively; many of the techniques articulated there can help you structure the plan. This roadmap exists in relation to your product roadmap. Some organizations run parallel tracks with allocated "lanes" for technical work versus product features (e.g., 70% product development, 30% technical development). Others opportunistically pull technical items into sprints when they align with product work. The key is being explicit about how technical and product work interweave, whether that's through dedicated technical sprints, percentage allocations, or by embedding technical work within feature development.

People, resources, and teams

Clearly describe how people and resources will be allocated, including team members, tools, and budgets. One aspect of your technical strategy may be team changes, such as growing a team, reorganizing a team, or redefining or redeploying people to better meet your needs.

Risk management and contingency planning

Identify and assess potential risks, including technical, financial, and operational factors. This is the "what could go wrong" section. You can't plan for every risk or plan contingencies for every potential risk, but you can likely identify high-level risks.

For example, you might identify a risk like the following:

Risk
> Our application relies on a third-party authentication service that has experienced three outages in the past six months, which has directly impacted our users' experience of the platform.

Mitigant
> By Q2, implement a fallback authentication method that can be activated within five minutes of detecting an outage in the primary service. Additionally, establish a monitoring system that alerts the on-call engineer when the service's response time exceeds two seconds. Over the longer term, consider a replacement for the platform.

Here is another example:

Risk
> We have a severe knowledge silo where only one engineer (Rajiv) understands our payment processing system, creating a significant bus factor problem.

Mitigant
> Schedule a weekly knowledge-sharing session where Rajiv walks another engineer through one payment system component. Document all sessions and store them in our knowledge base. By the end of Q3, ensure at least two additional engineers can confidently support the payment system.

Consider this technical debt example:

Risk
> Our monolithic codebase is becoming increasingly difficult to maintain, with bug fixes taking 30% longer than they did six months ago.

Mitigant
> Begin incremental migration to a microservices architecture, starting with the most frequently changed components. Set a goal to extract two services by the end of Q2, with clear interfaces between them and the remaining monolith.

Concrete examples make risk management actionable. Consider risks around scaling (e.g., if traffic increases 10 times), team changes (e.g., if your key architect leaves), vendor dependencies (e.g., if AWS has an extended outage), and technical evolution (e.g., if your framework becomes unsupported). The key is to be specific about both the risk (ideally with some quantification of its impact) and the concrete actions you will take to address it, including time frames and success criteria.

Governance and decision-making processes

Define decision-making processes and structures, specifying roles, responsibilities, and review procedures. Effective governance ensures alignment, consistency, and accountability throughout the project lifecycle. In Chapter 9, we'll discuss implementation governance when we introduce requests for comments (or RFCs).

Monitoring, metrics, and continuous improvement

Outline methods for tracking progress and measuring success. We discuss metrics more in Chapter 10, but every strategy needs qualitative and quantitative metrics to measure progress and success—or failure. It's also important to establish checkpoints, such as every two sprints or once a month, where you pause, digest your current state and the metrics, and determine what might need to be adjusted, improved, or terminated.

The Living Document Approach

The most critical thing to understand about your technical strategy document is that it's not meant to be static. Much like the system architecture diagram James created on his first day at the startup, which we discussed at the beginning of this chapter, your strategy document should evolve as your understanding deepens and conditions change.

We recommend treating it as a living document. We discussed scheduling regular reviews, either monthly or quarterly, which is a good cadence for most organizations. These reviews enable you to revisit the strategy, assess progress, and make updates based on what you've learned. This keeps it relevant and avoids the all-too-common fate of strategy documents: forgotten in some digital folder, never to be opened again.

However, creating a technical strategy document is just the beginning. Even the best-crafted strategy is worthless if it doesn't translate into actual change in how your team works and what you build. Let's discuss bringing your strategy to life in your organization's messy day-to-day reality.

Making Technical Strategies Real

So, you've got a slick strategy document—congrats! But as anyone who's ever created a PowerPoint deck knows, pretty slides don't magically translate into real-world change. Let's discuss how to make this happen in your organization's messy reality. Most strategies fail when they move from "Here's what we should do" to "We're actually doing it." You have to juggle multiple challenges at once—and it can feel like trying to solve a Rubik's Cube while riding a unicycle. We've identified some guiding principles for implementation that should help you keep on track, all of which tie back to the technical principles you've established.

Resource allocation (without burning everyone out)

Your most precious resource is your people. Be thoughtful about matching tasks with team members. Nobody wants to be the person assigned to "fix the legacy monolith" for the fourth quarter in a row. Spread the love (and the difficulties) around. Ensure people have the necessary tools and time, and don't expect them to work nights and weekends to meet arbitrary deadlines. That may work short-term, but you'll be writing new job ads before you know it. Your poor planning should not constitute an emergency for your team.

Picking tech that won't make future-you curse past-you

Choosing tech is like getting a tattoo: it'll be with you for a long time, or it'll cost you dearly to remove it, so choose wisely. We have already discussed "choosing boring technology" and not chasing shiny objects, but it's worth repeating. Ask yourself, "Will this solve our actual problems?" and "Can we maintain this when the person who championed it inevitably leaves?" Remember, you're not just solving for today; you're creating the foundation for whatever comes next. This directly implements your technical principles around simplicity, maintainability, and evolution.

Risk management (or, expecting the unexpected)

Something will go wrong—we promise. The question is whether you've thought about what those things might be and how you'll handle them. You don't need elaborate contingency plans for every possible scenario (that would be an overengineering red flag). Instead, identify your biggest risks, have a basic plan for each, and ensure everyone knows who is responsible for what when things get tough.

Regularly check in on your risks—new ones always pop up, especially in chaotic environments. Yesterday's "unlikely" might be today's "oh crap, that's happening right now."

"But that's not what I meant!"

Miscommunication and assumptions are the common sources of failure in technical programs. People use the same words to mean completely different things. What's "done" to an engineer might be "ready for testing" to a product manager and "ready to ship" to a customer success person. Or they make assumptions because something isn't transparent to them.

Create a shared language. Write things down. Have regular check-ins where people can ask "stupid" questions (spoiler: there really are no stupid questions). Document decisions and the reasoning behind them. The future you will thank you when remembering why you chose that particular approach six months ago. This implements your documentation principle in practice.

Keeping your finger on the pulse

We cannot emphasize enough that your strategy is a living thing, not a fossil. Set up simple ways to monitor how things are going. This could include dashboards, regular check-ins, or asking the right questions during your standups. The point is to know whether your brilliant plan is actually working before it's too late to make a course correction.

Look for both technical metrics (e.g., "Is the system performing better?") and human ones (e.g., "Is the team less stressed? Are we shipping faster?"). Both matter. This directly implements your "visibility matters" principle.

We cover metrics in more detail in Chapter 10.

Balancing all the things

Remember the trade-off discussion from earlier? Every technical decision involves them. Be explicit about what you're prioritizing and why. It's okay to say, "We're choosing this approach because speed to market is more important than perfect scalability right now." Just make sure everyone understands the trade-offs you're making. It's important that you document these decisions, as they embody your principles in action.

In Practice: April's Story

April faced a classic trade-off in the StudioOps team: the animation rendering engine needed optimization for 4K displays (customer request), but the authentication system needed security updates (compliance requirement). She made the trade-off explicit in a team meeting: "We can do 4K support in Q1 and security in Q2, risking a compliance audit failure, or we can do security first but potentially lose our biggest customer who needs 4K by March." The team discussed and decided on a hybrid: implement the critical security patches in two sprints (not the full rewrite), then focus on 4K, and then return to complete the security overhaul. April documented this decision in their architecture decision record: "We chose tactical security fixes over a complete authentication overhaul to balance compliance risk against customer retention. The full overhaul is scheduled for Q2 after the 4K feature ships."

Building capabilities, not just systems

Implementing new technologies often requires new skills. Ensure your team has the necessary training and support in place. This isn't just about formal training; it's about creating space for people to learn, experiment, and build confidence with new approaches.

In Practice: April's Story

When April standardized StudioOps on PostgreSQL, she didn't just mandate the change; she set up "PostgreSQL Fridays" where Derek (who knows PostgreSQL well) ran hands-on sessions. This included topics like "basic query optimization" in week 1, "JSON functions for document storage" in week 2, and "full-text search configuration" in week 3.

She also purchased a team license for a PostgreSQL performance course and allocated four hours per week for learning. The initial productivity dip was real—feature velocity dropped 30% in the first month. But by month two, the team was shipping faster than before because they weren't constantly context switching between database paradigms. The juniors especially benefited; instead of being confused by five different query languages, they could actually become proficient in one.

Be patient. Learning takes time, and productivity often dips before it improves when people adapt to new ways of working. That's normal and expected, not a sign of failure.

Communicating the vision

People need to understand what they're doing and why it matters. Help them by connecting the dots between individual work and the larger technical and business goals. Repeat the message more than necessary in every available medium; people need to hear something multiple times before it sinks in, and everyone consumes information differently.

And make it concrete. "We're improving system reliability" is vague. But "We're reducing customer-facing errors by 50% so people can complete their purchases without frustration" gives people something tangible to rally around. Your team should be able to understand the mission, know how it is being measured, and explain how it is happening, when it is happening, and what their involvement or ownership is.

Leading by example

Leaders play a crucial role in sustaining improvement. If you're in a leadership position, you need to champion the strategy visibly and show that you're open to feedback and adaptation. Nothing kills a strategy faster than leaders who say, "This is important," but then ignore it in their actions.

Celebrate wins, learn from setbacks, and connect the technical work to the larger mission. Help people see how their daily efforts contribute to the bigger picture.

The most potent form of communication isn't what you say; it's what you do. Leaders need to model the behaviors they want to see: embracing change, being open to feedback, and focusing on outcomes rather than perfect adherence to a plan.

If you say "learning from failure is important" but then publicly roast the first person whose project doesn't go as planned, guess what message people are actually hearing? Your actions always speak louder than your words. You must embody the technical principles you've established.

Conclusion: Enabling Execution with Technical Strategy

A good technical strategy aligns your decisions with your goals, strikes a balance between innovation and practicality, and sets you up for sustainable growth. It's not about creating the perfect plan but establishing a clear direction and approach that can evolve as conditions change.

The most successful strategies embrace the reality of change rather than trying to predict every detail up front. By building flexibility and learning into your approach, you become better equipped to navigate challenges.

Remember, at its heart, technical strategy is about enabling your organization to succeed, not just deploying cool technology. Keep your focus on the outcomes you're trying to achieve, be thoughtful about the trade-offs you're making, and bring your people along on the journey.

With a solid but adaptable strategy, you'll be better positioned to tackle the chaos, seize opportunities as they emerge, and build systems that can grow and evolve with your business.

In Chapter 9, we'll talk about how to take your strategy and turn it into technical practices.

Collaborative Technical Practices and Decision Making

In chaotic environments, the difference between success and failure often hinges on how effectively teams work together and make technical decisions. Without robust collaborative practices, even the most talented engineers can build the wrong things, creating unnecessary complexity or duplicating efforts—not to mention the inherent friction, discord, and conflict created by confusion and misdirection. For you to succeed, decision making needs to be crisp and fast, and everyone needs to move in the same direction quickly and safely.

Chaotic environments are already stressful, and if your team (and your leadership style) isn't in routine or doesn't support you, then working at a chaotic organization goes from being potentially stressful to untenable. In this chapter, we explore how to establish effective collaborative technical practices and decision-making frameworks that can thrive even in the most chaotic and turbulent organizations.

In Chapter 6, we discussed the dangers of processes for their own sake and the tendency of extensive, unwieldy processes that lead to inertia and inefficiency. We talked about being agile, starting with the smallest amount of process and then building on that, only if needed. The same advice applies to building collaboration and leveling up decision making. In our experience, people innovate in light-touch environments and they stagnate in high-touch environments.

We also introduced the concept of (shared) technical principles in Chapter 8. These will lead out this chapter, too. Like technical architecture, they make a good foundation for building good technical practices and making good decisions.

Shared Technical Principles

As we discussed in Chapter 8, technical principles serve as the foundation for engineering decisions, guiding both strategic planning and daily implementation. These principles aren't meant to be rigid rules but rather guides that help teams navigate complex technical landscapes. When developed collaboratively, they create a shared understanding of what "good" looks like in your organization, and they provide a framework for consistent decision making across teams and projects.

Also in Chapter 8, we articulated 14 proposed principles:

- Simplicity first
- Scalability
- Modularity
- Reliability
- Maintainability
- Documentation
- Testing

- Visibility
- Security by design
- Performance awareness
- Consistency
- Automation
- Openness
- Evolution

Importantly, for this chapter, you'll notice that almost none of the example technical principles can be interpreted as purely technical. Sure, most of them have a technical aspect, but they also include social or cultural elements; this isn't by accident.

Some of this is natural; principles tend to be more abstract than concepts like linting rules. You might build linting rules based on a technical principle. Still, it doesn't make much sense to specify tabs or spaces as a technical principle because of two main reasons:

- It's a very narrow focus.
- It doesn't deliver any real value to the team.

If you were to take that approach, your technical principles become overly detailed, and could start resembling a government paper:

> …section A, subsection 4, paragraph 7, clause 8, line 16 deals with the use of spaces (see section G.1.1.2 for definitions of "spaces" and space volumes) versus tabs (see section G.1.1.2 for definitions of "tab" as a spatial construct)…

And then they become impossible to reason about and, frankly, hinder innovation.

Other aspects are more deliberate consequences of articulating principles that focus on the intersection of people and technology. You see, technical principles do double duty: they guide technical implementation and shape your team's culture and social dynamics.

Take "simplicity first," for example. On the surface, this principle is about writing cleaner code with fewer moving parts. But in practice, it transforms how your team operates on a day-to-day basis. When an engineer starts suggesting an elaborate, 12-microservice solution to a problem that could be solved with a simple API endpoint, "simplicity first" gives the team permission to pump the brakes and ask, "Do we need all that complexity?"

"Simplicity first" also doesn't just become a default response to scope creep. In planning meetings, when requirements start ballooning, the team can say, "Let's simplify first," and refocus on the essential problem. This principle then shapes not just architecture but the entire approach to work—from how you run meetings (shorter, more focused) to how you onboard new team members—starting with core concepts before diving into details.

Or consider the principle of "documentation." Beyond ensuring your code has comments, this principle cultivates knowledge-sharing behaviors. Teams that value documentation tend to develop cultures where teaching others is celebrated. They're more collaborative because they understand that undocumented knowledge creates dangerous dependencies on specific individuals.

The beauty of technical principles is that they create a shared language that bridges technical implementation and human behavior. When your team invokes the "automation" principle, they're not just talking about writing infrastructure as code; they're reinforcing a culture that values everyone's time and focuses human energy on creative problem-solving rather than repetitive tasks.

These principles should become cultural touchstones, invoked in meetings and Slack threads. They help new team members understand how things are done far more effectively than any employee handbook. They're powerful precisely because they operate at the intersection of technology and people, guiding both what you build and how you work together to build it.

In Practice: April's Story

At PixelCurl, April saw this principle-culture connection firsthand. When she arrived, there were no shared principles—just three different philosophies from three senior engineers. Marcus believed in "performance above all," optimizing every millisecond even if it meant unreadable code. Sarah championed "move fast and break things," shipping features quickly but leaving a trail of bugs. Derek practiced "fortress engineering," building bulletproof systems that took months to modify. But this resulted in junior engineers getting whiplash trying to follow three contradictory approaches. So, April facilitated a principles workshop where they arrived at their first shared principle: "simplicity first." Within weeks, code reviews became less combative. Marcus stopped writing incomprehensible one-liners that saved 2 ms but cost hours to debug, Sarah started adding error handling instead of assuming happy paths, and Derek began accepting "good enough" solutions that could ship in the correct quarter instead of the next year.

Making Technical Principles Stick

Making technical principles stick requires visibility and regular reinforcement. Rather than tucking them away in a forgotten document or slide deck, you and your team (and this is ideal work for staff and principal engineers to wield influence) need to prominently display principles and reference them regularly during design discussions, code reviews, and planning sessions. We've seen organizations that printed their technical principles on the wall of their main workspace. For James, who is somewhat of a cynic, it initially seemed theatrical. Still, he's come to appreciate that there is a power in having principles be a natural part of vocabulary.

Mission statements, vision statements, and technical principles only become items without power if they are treated as props or parroted as rhetoric. If they are treated with respect and reinforced, then they become touchstones that push people to do better. In chaotic environments, we've learned that morale, team unity, and focus are fragile things; cultural touchpoints like this reinforce that unity and continuity.

Perhaps even more important than displaying principles is modeling them through leadership behavior. If you preach simplicity but approve overly complex designs, or champion documentation while leaving your code undocumented, your team will follow your actions, not your words. The disconnect between stated principles and leadership behavior creates cynicism and undermines the entire framework. Technical leaders must demonstrate their commitment to these principles through their work and decisions, creating a culture where principles are lived rather than just spoken.

Technical principles should evolve as your organization changes. Schedule regular reviews, perhaps quarterly, to assess whether your principles still align with your team's goals and challenges. This keeps them relevant and reinforces their importance. The review process itself can be a valuable opportunity for team discussion, helping newer team members understand the rationale behind each principle and enabling everyone to contribute to their refinement. As your team grows and technology evolves, your principles may need to adapt while still preserving their core intent.

Communication, Collaboration, and Execution

Effective communication is essential for collaboration, yet teams frequently face obstacles in this area. Nowhere is this more obvious than in shipping products. Imagine your team's product (and its requirements, codebase, and execution) as a shared kitchen where everyone's cooking different dishes simultaneously. When engineers are all working in the same space without talking to each other, then it's hard to reason about what to cook, how many ingredients you need, and where your kitchen utensils are. This kind of uncoordinated effort leads to confusion, duplicate work, and sometimes even a few burnt meals.

So, how do you solve this? The most common natural reaction is to enforce collaboration and communication through rigid structure and centralized command and control. We can extend our kitchen metaphor to provide an exploration of why this doesn't work. Professional kitchens rely on the brigade system. In this well-defined structure, everyone has their station and clear responsibilities. The head chef (tech lead) coordinates the overall vision, while station chefs (senior engineers) oversee specific areas, such as sauces, meats, or desserts (databases, frontend, APIs). Less experienced chefs work within these stations, learning the craft while staying in their assigned area. The magic allegedly happens because everyone knows exactly where their responsibilities begin and end, communication follows established patterns ("Yes, chef!"), and the implication is that the structure allows for both specialization and smooth coordination.

Ah, no. Both James and Juan worked in the hospitality industry. Indeed, Juan was a professional chef before turning to the dark side, and the brigade system is a good example of how *not* to collaborate to ship products. The strict hierarchies and rigid role definitions of the brigade system often fail when building products. The brigade system was designed for execution and repetition, not creation. A restaurant kitchen's primary goal is to repeatedly produce the same dishes to specification, night after night. The head chef designs the menu, and everyone else executes it with military precision. But product development isn't about executing a fixed plan; it's about discovery, iteration, and problem-solving in environments where requirements constantly evolve.

The brigade system's strict hierarchy stifles the very things that product teams need most: cross-functional collaboration, creative problem-solving, and rapid adaptation.

In a traditional kitchen, a line cook who suggests improvements to a dish might be told, "Just cook it as directed." But that mindset is deadly in product development, where the best solutions often come from unexpected places and from the team members who are closest to specific problems.

The knowledge silos created by strict station assignments are another major issue. In a brigade, the saucier might not know much about pastry, and that's considered acceptable. But in product development, when a backend engineer is the only one who understands the data model, or a frontend engineer is the only one who knows the UI framework, you've created dangerous dependencies and bottlenecks. When that person gets sick, goes on vacation, or leaves, the entire product suffers.

In Practice: April's Story

April discovered the studio operations team's communication breakdown through a critical incident. The authentication service (Marcus's domain) failed at 2:00 a.m. This resulted in Derek and Sarah spending four hours debugging because Marcus hadn't documented the custom JWT rotation scheme that he implemented. When Marcus returned from vacation, he fixed it in five minutes. The post-assessment revealed similar patterns everywhere: Sarah's React components used an undocumented state management pattern that only she understood, Derek's database had stored procedures with cryptic names like "sp_process_thing_v2_final_FINAL." So, April implemented "Documentation Fridays," where the last two hours of every Friday are dedicated to documenting one piece of institutional knowledge. She made it stick by having each senior engineer present their documentation to the juniors, who asked questions until they could operate the systems themselves.

The command-and-control communication style is the most problematic aspect. "Yes, chef!" might keep a dinner service moving, but it doesn't encourage the questioning and challenging that are necessary for innovation. James has seen countless product teams fail because they adopted this top-down approach, where decisions flowed from leadership without meaningful input from those doing the implementation work.

Modern product development thrives on cross-functional teams where boundaries blur, roles overlap, and everyone contributes to both problem definition and solution. Engineers need to understand user needs, designers need to grasp technical constraints, and product managers need to appreciate both. The brigade's neat separation of duties simply doesn't work when building complex products in uncertain environments.

We discussed some of these ways of working in Chapter 6.

So, while there's certainly value in clear responsibilities and coordination, successful product teams look less like military brigades and more like jazz ensembles; individual expertise combined with collaborative improvisation, guided by shared objectives rather than rigid hierarchies. They maintain enough structure to stay coordinated but remain fluid enough to respond to the changing needs of users and markets.

Saying that this is easy is one thing, but getting there is much harder. So let's talk about how you can achieve this.

Communicate, Communicate, Communicate

Communication forms the backbone of effective technical collaboration. In chaotic environments, where requirements shift and priorities change daily, getting communication right is what separates teams that excel from teams that struggle. As we discussed in Chapter 3, teams must develop deliberate communication strategies. These strategies need to account for the complex nature of technical work, the informational density of many of the concepts, and ensure that a diverse community of cross-functional collaborators all get on the same page.

Building software can be incredibly complex, which means we have a wide range of potential communication pitfalls and barriers that can appear—sometimes from multiple sources—at the same time. Technical jargon creates walls between specialists and generalists. Distributed teams struggle with time zone differences and the loss of nonverbal cues. High-pressure, chaotic environments can lead to terse, easily misinterpreted messages. Addressing these barriers requires deliberate effort to create communication norms that work for the team.

As we've discovered, communication isn't just about transferring information; it's about creating a shared context. As your team and products grow, the risk of context loss increases dramatically. New team members join without the historical understanding that guided initial decisions. Engineers can become siloed and focus intensely on their domains without seeing broader connections. This fragmentation of knowledge leads to misaligned efforts and inefficient solutions. Great technical leaders—both engineering leaders and individual contributors—recognize that their job isn't just to share information but to create a coherent narrative that connects individual contributions to the larger mission.

Cadences

So, we start by establishing clear communication cadences that help maintain alignment without drowning teams in meetings. Daily check-ins provide quick course corrections, weekly syncs enable deeper problem-solving, and monthly reviews connect technical work to broader business objectives. These structured touchpoints create predictable information flows, reducing uncertainty and helping teams stay coordinated without constant interruptions.

It's hard to provide a calendar or agenda for good cadences, as they vary from team to team and are mediated by time zones and geographies. What we can tell you is that if you poll your team and ask "Do you know what you need to know?" and the answer is "No" or just a look of panic, then looking at how frequently you share information is a good starting point to solve the issue.

Mediums and modalities

We also need to be flexible about the medium or mediums of these communications. James has witnessed this firsthand, repeatedly, when leading distributed teams across multiple continents. Initially, teams tend to rely primarily on one form of communication: we meet weekly via video conference or write documentation. However, time zones, cultural differences, and individual preferences in communication styles quickly lead to frequent misunderstandings and gaps. The only way to fix this is by establishing a multimodal approach—for example, using synchronous video for complex discussions, asynchronous audio messages for nuanced explanations (everyone loves a podcast!), and written documents for reference. This will dramatically reduce communication failures. The key is recognizing the following:

- Different types of information flow more effectively through different channels.
- Different types of information are consumed differently by different people and teams.
- Different types of information are more sensitive to external pressures, like time zones, language, and cultural differences.

So, we need to create norms that match the medium to the message. For technical teams, in particular, choosing the right communication tools can make all the difference. Architecture discussions benefit from visual collaboration tools with real-time diagramming capabilities. Design, architecture, or API contract negotiations work best with structured documentation tools that track changes and discussions. Day-to-day coordination might happen in persistent chat tools with organized threads and searchable history. The most successful teams don't just pick tools haphazardly; they thoughtfully match communication channels to the types of information being exchanged. They also multicast the information via multiple mediums.

But you're going to spam people! Well, no—spamming people is sending the same information multiple times via the same medium. "I sent the email six times," is pretty much guaranteed to result in someone ignoring your email. But sending an email, a Slack update, updating a Notion document, and reiterating in a video conference are not spamming, but rather multicasting.

Visualizations

While the argument that most humans are visual learners is probably false, there's no denying that visual information can dramatically speed up understanding of complex technical concepts. Visuals work because they bypass the limitations of serial processing in text and speech. Our brains process visual information holistically and in parallel, making it easier to grasp relationships, hierarchies, and patterns. That's why an exemplary architecture diagram can convey more meaning than pages of explanation, and why whiteboarding remains the go-to activity when technical folk get stuck.

But not all visualizations are created equal. Compelling technical visuals strike a balance between simplicity and accuracy. James once worked with a group of architects whose diagrams were so detailed and comprehensive that they became as impenetrable as the systems they tried to explain. The best technical visualizations follow the principle of progressive disclosure—showing the big picture first, and then enabling viewers to drill down into details as needed.

Consider using tools like Miro, Figma, and LucidChart, or hybrid tools like Mermaid, that enable you to embed visuals in other tools. These, or indeed any drawing tool, have transformed how teams create and share visuals. The collaborative aspects of these platforms mean that diagrams become living artifacts that teams build together rather than static images created by a single person. When visuals become collaborative, they better represent a shared understanding than individual perspectives.

Standardize your visual language across the team. Whether you're using Unified Modeling Language (UML), C4, or your custom notation, consistency helps everyone interpret diagrams correctly. Also, be consistent in using colors. And don't forget that ~8% of the population has color vision deficiency (*https://oreil.ly/putXQ*).

Video calls

The pandemic normalized video calls as a primary communication channel, but most teams still haven't figured out how to make them truly effective. Video calls enable you to convey emotion, build relationships, and handle complex discussions where facial expressions and tone matter. They're far from perfect, but they represent the best possible substitute for real-time, in-person interaction. They are, however, broadly terrible for detailed technical work and deep problem-solving unless carefully structured. Slides deserve a special callout here. For most video calls, slides render all the benefits of interaction—however limited they are—completely worthless. Despite the temptation, avoid slides. Please.

High-performing technical teams approach video differently. They use video strategically, rather than as a default, and they structure calls to take advantage of the medium's strengths. For quick status updates or information sharing, asynchronous tools are usually better; save video for conversations that benefit from real-time interaction and emotional connection.

In the past, we've seen people use video to build relationships across globally distributed engineering organizations. For example, when James was at Kickstarter, an app called "Coffee Roulette" was developed that matched people across the organization who didn't work together or might not have met. The company paid for them to have a (physical) coffee together. This concept can be extended to be global and remote: "virtual coffee" sessions with optional 30-minute video calls. The company buys you a pastry and coffee or the like, and off you go—no agenda beyond getting to know each other.

For technical discussions, the most effective video calls incorporate shared visual spaces. Tools like meeting apps with whiteboard features, screen sharing with drawing capabilities, or integration with collaborative design tools can transform passive listening into active collaboration.

Most importantly, great teams establish explicit video call norms. Camera expectations, muting practices, and participation signals can reduce the cognitive overhead of video interaction. For example, raising hands for attention or to show reactions helps keep conversations flowing and limits the potential for derailment.

Audio

Audio remains the unsung hero of technical communication. While video dominates synchronous remote interaction and text dominates asynchronous exchanges, audio offers some cool benefits that we suggest you explore.

Voice messages deliver nuance and emotion more effectively than text without requiring the scheduling and full attention that video calls demand. An interesting potential use is asynchronous voice messages for code reviews that involve significant feedback. When an engineer needs to explain why a particular approach might cause issues, a two-minute voice note conveys both technical concerns and a supportive tone in a way that text comments often fail to do.

Recorded technical discussions create valuable learning resources with minimal additional effort. Simply recording architecture discussions or technical planning sessions creates an archive that new team members can consume at their own pace. With

AI and transcription you can also turn these recordings into documents or even summarized into an architectural design record (ADR). Some teams even record pair-programming sessions on key components or addressing common challenges in a codebase—an invaluable resource for onboarding that requires virtually no extra work.

Podcasts and audio lunch-and-learns offer accessible, continuous learning opportunities. One engineer whom James worked with started recording short technical explainers about their system architecture while driving to work. Combined with a few simple hand-drawn sketches, these ended up as supplements to the documentation.

Audio's biggest strength might be its compatibility with other activities. Unlike reading or watching a video, which demands visual attention, audio can be consumed while doing other things like taking a walk or making lunch. This makes it particularly valuable for teams feeling overwhelmed by information; audio offers a way to keep learning and stay connected without adding more screen time to already overloaded days. When James feels overwhelmed with activities, he often breaks, leaves his office, and listens to a recorded meeting that he couldn't attend (though normal people would or should just listen to a podcast).

Chat and messaging

Chat tools like Slack, Discord, and (sigh) even Microsoft Teams have transformed day-to-day technical communication, creating persistent and searchable conversation spaces that blend synchronous and asynchronous interactions. When used thoughtfully, these platforms dramatically reduce email overload and create more inclusive conversations. When used poorly, they become overwhelming firehoses of disorganized information.

Effective teams create intentional structure in their chat environments. Rather than having general channels that become noise factories, they create purpose-built spaces with explicit scopes. And they create those spaces iteratively. If you don't need a channel yet, then don't create one until it becomes clear that the conversation needs its own home. Juan is a particular exponent of carefully named channels using prefixes:

Team-
    For channels for specific teams

Proj-
    For particular projects

Ext-
    For channels with external parties

Well-structured threads keep conversations coherent without splintering team awareness. Teams with strong threading habits maintained much more transparent communication even as they scaled. The simple practice of responding in threads rather than the main channel prevented conversations from overlapping while preserving context.

Chat also creates valuable opportunities for personality and team culture to shine through. Emoji reactions, GIFs, and informal banter might seem trivial, but they build essential social bonds in distributed teams. Both James and Juan like to establish channels like a pets channel, a chatter channel where people can say hi and talk about their day, and a channel for new folk to introduce themselves. There is even a "weird-Australian-s***" channel in one of Juan and James' Slack channels, where the small community of Australians in the team share the weird and wonderful aspects of Australia that seem to bewilder and charm Americans (and yes, the spiders are that big).

Most importantly, high-functioning teams establish healthy expectations around chat response times and availability. The always-on nature of messaging platforms can create unhealthy pressure without explicit norms.

Email

Email might feel like (is?) the dinosaur of technical communication tools, but it remains surprisingly resilient for good reasons. Unlike the constant stream of chat or the ephemeral nature of conversations, email creates a persistent, searchable record with built-in threading that works across organizational boundaries. Despite predictions of its demise, email continues to serve vital functions in the communication ecosystems of technical teams.

Where email truly shines is in communications that need formality, a paper trail, or external reach. Email remains the universal connector across organizational boundaries, making it invaluable for technical discussions that extend beyond your immediate team.

Email's asynchronous nature and reduced urgency expectations make it well suited for thoughtful, nonurgent communication. We've found that technical discussions in email tend to be more carefully considered than the rapid-fire exchanges in chat tools. For complex technical proposals or detailed feedback that requires careful wording, email provides space for reflection that faster-paced channels don't encourage. Email also has numerous automation shortcuts: categories, tags/labels, and folders, and most modern email suites have the ability to decorate and route emails using their source, destination, subject, etc.

However, email's weaknesses are well-known and significant. Its one-to-many broadcasting nature can create overwhelming inbox volume when misused. Teams that default to "Reply all" for routine acknowledgments often find themselves overwhelmed with low-value messages. The same applies to notifications from code reviews or observability systems—too many of them, and alert fatigue sets in, causing emails to be ignored.

Another common pitfall is using email for real-time coordination or discussions that require rapid iteration. Email's inherent latency makes it poorly suited for dynamic conversations. If your email thread is 10+ emails long, then maybe a different medium is needed. Recognizing when to move conversations to more appropriate channels is a crucial skill for effective communication.

The most successful teams view email as a specialized tool rather than a default. They establish clear guidelines about what belongs in email versus chat, documentation systems, or synchronous conversations. Despite newer tools with flashier features, email's staying power comes from its universality, flexibility, and familiarity. Rather than trying to eliminate it, aim to thoughtfully integrate email into the ecosystem, leveraging its strengths while compensating for its weaknesses with other channels.

In-person interactions

Despite the rise of remote work, in-person interaction remains uniquely powerful for certain types of technical communication. The richness of face-to-face discussion— with its natural turn-taking, subtle nonverbal cues, and shared physical context— makes it especially valuable for complex problem-solving and building relationships. In our humble opinion, whiteboarding sessions continue to outperform digital equivalents for fluid ideation and system design. The combination of an ample visual space, physical movement, and immediate feedback loops creates a creative environment that is difficult to replicate virtually. Even teams that are mostly remote will find tremendous value in occasional in-person architecture or design sessions.

Pair programming takes on new dimensions when two engineers share physical space. The reduced friction in switching who's typing, the ease of pointing at specific parts of the screen, and the natural conversation flow make in-person pairing particularly effective for knowledge transfer and tackling thorny bugs.

Informal interactions around the office create invaluable serendipitous knowledge sharing. The classic "overheard conversation at the coffee machine" that solves a problem someone didn't even know they had might seem like a cliché. Still, these moments genuinely build team cohesion and cross-pollinate ideas.

For distributed teams that can't meet in person regularly, off-site gatherings become even more crucial. Targeted team gatherings every few months provide the foundation for relationship building that makes all other communication channels more effective. The most successful off-sites will balance structured technical work with unstructured social time, recognizing that both contribute to better collaboration.

Documentation

Documentation plays a critical role in technical communication, serving as an external memory for the team. But documentation is only valuable if it's read and used. Let's face it—great code with terrible documentation isn't great code; it's a future nightmare waiting to happen. The moment your star engineer wins the lottery and moves to Bali, that undocumented authentication service becomes a ticking time bomb.

Successful teams view documentation not as a tedious checkbox exercise, but as a form of time travel—with messages sent to future team members, including their future selves. James once worked with a team that had a security incident, and then had to reverse-engineer their API to fix it because the original engineers had moved on without documenting it. What should have been a quick intervention took three weeks of archaeological code excavation and iterative change as new things were learned and discoveries made. After this painful experience, they adopted a "documentation as you go" approach (and wrote tests) because no one wanted to go through that experience again.

Different types of documentation serve different purposes, and recognizing these distinctions helps teams create more valuable resources. Reference documentation—such as API specs, function signatures, and database schemas—benefits from automation and standardization. Conceptual documentation (e.g., architecture overviews, design rationales, and mental models) requires thoughtful human explanation accompanied by plenty of visuals. Procedural documentation (setup instructions, troubleshooting guides, operational runbooks) needs clear, step-by-step sequences with examples. Teams that try to treat all documentation the same way often end up with resources that serve no purpose.

The perennial challenge is keeping documentation fresh as systems evolve. Static documentation inevitably drifts from reality, sometimes becoming worse than no documentation at all by leading engineers down incorrect paths. We're big fans of collocating documentation and code. Moving critical documentation directly into the codebase and into places like pull request templates puts it front and center. We also recommend including documentation updates into the definition of "done"; if the docs are not updated, then the team can't claim that they have completed the fix or feature. The best documentation lives where the work happens, whether that's in code comments, README files, or tools that engineers use daily.

Effective documentation strikes a balance between comprehensiveness and conciseness. The sweet spot is documentation that answers the most essential questions clearly and points to where more details can be found if needed. There is an adage (that neither of us can remember the origin of): "Document the 'why' thoroughly, the 'what' clearly, and the 'how' only when it's not obvious from the code."

The best teams establish clear ownership of documentation without making it the sole responsibility of one person. Making documentation enjoyable to create and consume pays enormous dividends. Teams that view documentation as a creative challenge rather than a burden tend to produce resources that get used.

Finally, a documentation culture requires leadership investment and setting an example. When senior engineers and architects treat documentation as optional or low priority, then junior team members will follow suit. When leaders explicitly value clear explanation alongside working code—by celebrating good documentation, allocating time for creating it, and visibly using it themselves—they establish norms that strengthen the team's collective knowledge base.

Confirming understanding and handling disagreement

Communication isn't complete until the message is received and understood, not just sent. In technical teams, this gap between transmission and reception causes countless problems: features built to the wrong specification, architectural decisions that surprise stakeholders, and teams working at cross-purposes despite regular meetings.

The first challenge is verification. How do you know that your message landed? Active confirmation techniques can help bridge this gap. After explaining a complex technical concept, ask team members to summarize it back in their own words. This isn't meant to be patronizing; frame it as ensuring that you explained it well: "Let me check I've explained this clearly; how would you describe what we're building to someone else?" For written communications, request explicit acknowledgment of key decisions or action items. A simple "Please confirm you're aligned with this approach" prevents the silent disagreement that surfaces weeks later as rework.

When people genuinely don't understand despite multiple attempts, the problem usually isn't their comprehension; it's your explanation. Try switching modalities. If verbal explanation isn't working, try visual diagrams. If documentation isn't clicking, record a video walkthrough. Different people process information differently, and flexibility in your communication approach shows respect for these differences.

Disagreement requires even more nuance. There's passive disagreement, where someone nods along but never truly buys in, and active disagreement, where concerns are voiced openly. Passive disagreement is more dangerous; it manifests as slow execution, subtle sabotage, or surprise objections late in the process. Create explicit opportunities for dissent: "What concerns do you have about this approach?" or "If you had to argue against this, what would you say?" Making disagreement safe and expected brings it into the open where it can be addressed.

When facing active disagreement, resist the urge to immediately counterargue. First, ensure you understand the objection completely. Repeat it back: "So your concern is that this approach won't scale beyond 10,000 users; is that right?" Often, the act of clarification reveals that you're not as far apart as initially thought. If genuine disagreement remains, acknowledge it explicitly and document it. "Sarah has raised concerns about scalability. We're proceeding with this approach because we need to ship quickly, but we'll revisit if user growth exceeds projections." This shows you heard the concern and creates a clear trigger for reassessment.

Sometimes, despite your best efforts, the message simply won't land. After three failed attempts using different approaches, it's time to escalate or find a translator—someone who can bridge the communication gap. This isn't failure; it's pragmatic recognition that sometimes a fresh voice or different relationship dynamic is what's needed to break through.

Transparency and ambiguity

One of the primary challenges in technical communication is the inherent ambiguity that often characterizes chaotic settings. Teams must adapt to unclear situations while making purposeful decisions. This requires a careful balance of empowering team members to express their priorities and ensuring that decisions align with the overall project goals. Leadership plays a crucial role in shaping team dynamics and facilitating a culture that encourages collaborative decision making without imposing top-down directives.

Ambiguity creates anxiety, and anxious teams tend to make poor decisions. Leaders can reduce this anxiety, not by pretending to have certainty where none exists, but by creating frameworks that help teams navigate uncertainty productively. This might include establishing transparent decision-making processes, defining what "good enough" looks like for different types of decisions, and creating spaces for structured speculation about possible futures.

Juan is fond of saying, "Default to open and public." He can often be found lurking in Slack group chats, asking why this isn't in a public channel. Creating a culture of transparent communication means recognizing that information is power, and that power should be distributed throughout the organization. When technical decisions happen behind closed doors or without proper context, teams feel disempowered and disconnected from the outcomes. By contrast, organizations that share information broadly—including the reasoning behind decisions, not just the decisions themselves—build trust and enable better autonomous decision making at all levels.

Ultimately, effective communication in technical teams isn't about perfection; it's about creating an environment where misunderstandings are quickly identified and resolved. The best teams develop a shared language, establish explicit communication norms, and foster a culture where asking questions and seeking clarification are valued rather than penalized. With these foundations in place, even the most chaotic environments become navigable, enabling teams to collaborate effectively and deliver exceptional products.

Exercise: The Architecture Telephone Game

Finally, here's an exercise we like that's a fun activity focusing on communication, documentation, knowledge transfer, and technical precision. It's also a pretty cool team-building exercise for an off-site. Follow these steps:

1. Have your team sit in a circle, whether in person or virtually.

2. Select a moderately complex system or feature that your team has built or is familiar with.

3. Select an initial participant. They draw (or better, use an existing diagram) a high-level architecture diagram of this system on paper or a digital whiteboard, visible only to them.

4. Select another person, and then, in another room or breakout room, the first participant has two minutes to verbally describe the architecture to the next person without showing them their diagram. This second person draws what they understand from the description.

5. You then bring in a third person, and continue the process around the circle.

6. After reaching the last person, compare everyone's diagrams side by side. The differences between the first and last diagrams often reveal fascinating communication gaps and assumptions. Discuss specific moments where information was lost or transformed. Was it terminology differences, missing context, implicit knowledge that wasn't stated, or something else?

The real value of this exercise comes in the debrief, where you identify patterns in how technical information degrades across transmission points. Teams who practice this regularly develop more precise technical vocabulary, learn to verify understanding through targeted questions, and become more aware of when they're using jargon or making assumptions. Also, some of the final drawings are often pretty funny.

Collaboration

Cross-functional communication and collaboration aren't just about having the right tools; it's about creating a culture where different disciplines actively seek each other's input and feedback. This requires deliberate effort to break down "us and them" mentalities that can develop between engineering, product, design, and other functions. Successful teams create opportunities for meaningful interaction, from informal coffee chats to structured cross-disciplinary pairing sessions.

Effective cross-functional collaboration requires structural support. This might include tools, shared goals and metrics, cross-disciplinary habits like joint planning sessions, and leadership that rewards collaborative behaviors rather than siloed excellence. The goal isn't to eliminate specialization but to ensure that specialists work together effectively toward common outcomes.

Like bringing Slack conversations into open channels, the first step to collaboration is ensuring that everyone who needs to be involved in a decision or discussion is included and takes ownership of the outcome. If everyone has a stake in the outcome, then they are more likely to work together to resolve it. This also reduces the incidence of the "throw it over the wall" mentality.

But breaking down this notorious mentality takes persistent effort. James once worked with a team where engineers built features exactly as specified, only to have them rejected by users. So, what was the root cause? Product managers were writing requirements without engineering input on feasibility, and engineers were implementing without asking clarifying questions. To solve this, James instituted kickoff sessions where product, engineering, and quality assurance teams collaboratively refined requirements before work began.

The physical (or virtual) environment has a massive impact on collaboration quality. Again, the default should be to open: have project and team channels where all parties can gather. Frequently, we see the engineering channel in Slack being private. Why? "So nonengineers don't bug us" is usually the answer. Putting aside that engineers have jobs because people who bug them want things, those who ask in a public channel can be managed by not removing access to the public channel. Try directing people to a support channel instead of hiding your work from others.

Remote teams should be intentional about creating opportunities for serendipitous collaboration that might naturally occur in colocated environments. Some organizations implement virtual "office hours" where technical leaders or domain experts make themselves available for impromptu discussions and questions. Others create dedicated channels for specific technical domains or projects, encouraging ongoing discussion and ideation. These informal collaboration spaces complement more structured decision-making processes by enabling ideas to emerge and evolve through casual interaction before they are formalized into proposals.

Visual collaboration tools become critical for remote technical discussions that would traditionally happen at a whiteboard. Shared diagramming applications, digital Kanban boards, and collaborative documents enable distributed teams to create and manipulate shared representations of technical concepts in real-time or asynchronously. Teams that invest in building fluency with these tools and establishing norms around their use can maintain high-quality technical discussions despite physical distance. The most effective remote teams treat their virtual collaboration environment as a priority concern, continuously refining their tool choices and practices to support rich technical communication across distances.

Shared language is also key for effective collaboration. When engineers talk about refactoring and technical debt, while product folk discuss user journeys and market fit, miscommunications are inevitable. This extends to explaining technical jargon and concepts to nontechnical audiences. Building a sense of empathy and understanding across functions will pay enormous dividends. Teams that understand each other's challenges and constraints make better decisions together. Celebrating cross-functional wins specifically—not just team or individual achievements—reinforces collaborative behaviors.

Psychological safety becomes even more crucial in cross-functional settings. When people fear looking ignorant in front of colleagues from other disciplines, they won't ask clarifying questions or offer alternative perspectives. One practice that works wonders is "designated novice" questions—having senior people deliberately ask basic questions that others might be afraid to voice. James worked with a senior engineer who used this technique regularly to tease out and encourage questions by asking, "Can you explain that concept again? I'm not following." It created permission for everyone else to seek clarity without embarrassment.

There's more on psychological safety in "Psychological Safety" on page 201.

Most importantly, true collaboration means sometimes changing direction based on input, not just collecting opinions and proceeding with your original plan. Nothing stifles the collaborative spirit faster than the perception that "my input never matters." Openly acknowledging the value of that input and changing course creates a culture where people know that their expertise genuinely influences outcomes. This strengthens collaborative bonds for all future work.

Remember, collaboration isn't some corporate feel-good exercise; it's the practical recognition that complex products require diverse expertise working in concert. When engineers understand user needs, designers grasp technical constraints, and product folk appreciate implementation realities, that's when the magic happens. The result isn't just better products but more engaging work for everyone involved.

For James, at a large organization with many independent and remote teams, there were a lot of handovers from one team to the next: product to design, design to engineering, engineering to quality assurance. There was cognitive overhead in the handovers, and the teams frequently were not on the same page about requirements or timelines. James proposed a trial of cross-functional teams, and put everyone working on a specific product or feature in a single pod. Almost overnight, the level of collaboration soared. Instead of handovers, conversations happened in standups and everyone had a clear picture of requirements and expectations. With collaboration came engagement, improved execution and delivery. It was a simple change with a big impact.

Execution

Finally, there's execution: the tasks and activities that make up day-to-day work for engineering teams. We're going to touch on some of the key practices that, when done correctly, will pay dividends.

Effective coding practices

While technical principles provide high-level guidance, coding practices translate these principles into concrete, everyday behaviors. These practices cover everything from naming conventions and testing requirements to architectural patterns and documentation standards. They're the practical manifestation of your principles; the "how" that supports the "why" of your technical vision.

The most effective coding practices emerge collaboratively rather than being imposed from above. When engineers participate in defining standards, they're more likely to follow them and advocate for their adoption across the team. This doesn't mean decisions being made by a committee—someone still needs to make the final calls—but the process should incorporate diverse perspectives and experiences. Collaborative development of practices creates buy-in and ensures the standards address real problems the team faces rather than theoretical concerns or personal preferences of a single senior engineer.

Consistency is the primary goal of coding practices. A consistent codebase is easier to understand, maintain, and extend, especially for new team members. But achieving consistency requires more than just writing down rules; it requires reinforcement through mentorship and review processes. Regular code reviews provide a powerful mechanism for enforcing coding practices while simultaneously serving as learning opportunities. When done well, they help disseminate knowledge throughout the team and elevate everyone's skills. The key is approaching reviews with a spirit of collaboration rather than criticism, focusing on the code rather than the coder.

Like technical principles, coding practices should evolve with your team and technology. What works for a team of three might not work for a team of 30. Regular retrospectives can help identify which practices are adding value and which need refinement.

Fortunately, you don't need to create coding practices from scratch. Many open source projects and technology companies publicly share their standards, enabling you to steal—err—providing excellent starting points that you can adapt for your team.

Navigating code-review drama

It's important to directly and specifically call out code reviews. Before diving into the interpersonal dynamics of code reviews, it's worth establishing why we do them at all. Code reviews serve multiple critical purposes: they catch bugs before they reach production, ensure code meets team standards, spread knowledge across the team, and provide mentorship opportunities. Most importantly, code reviews are about improving the collective output of the team, not gatekeeping or demonstrating superiority. When everyone understands that reviews exist to make the codebase better and help each other grow, the emotional temperature drops significantly. The goal isn't to find fault; it's to collaboratively ensure that what ships is maintainable, understandable, and aligned with your technical principles.

Code reviews can be tense if not handled carefully. Reviews that are too harsh can demoralize team members and create a culture of fear, while reviews that are too lenient can let quality issues slip through and accumulate technical debt. Finding the right balance requires empathy, clear expectations, and a shared understanding of the purpose of reviews.

Consider this common scenario: a junior engineer submits their first pull request, only to receive a barrage of critical comments from a senior engineer. Feeling discouraged, they approach another senior engineer, asking if they're a terrible engineer who should quit. This situation requires addressing both immediate and systemic issues. In the short term, you need to reassure the junior engineer that critical feedback is part of the learning process, not a judgment of their worth. Help them understand that even experienced engineers receive substantial feedback on their code and that the review process is about improving the codebase, not criticizing the engineers.

In Practice: April's Story

April faced this exact scenario in her second week at PixelCurl. Emma, a junior engineer, submitted a pull request for a new API endpoint. Marcus left 47 comments, including gems like "This is wrong" and "Did you even read the documentation?" Emma went to April in tears, convinced she's not cut out for engineering. So, April reviewed the pull request herself. She found that Emma's code was actually solid, just not following Marcus's personal preferences (which weren't documented anywhere). April implemented a three-part solution. First, she created a code review checklist distinguishing "must fix" (security issues, bugs) from "consider" (style preferences). Second, she made it clear that tone matters; things can be critiqued but every critique must be constructive. Third, she paired Marcus with Emma for a week, instructing him to explain his standards in person. Marcus discovered that Emma has clever insights when not terrified, and Emma learned that the way Marcus critiques stems from caring about quality, not personal attacks.

At the same time, it's important to speak privately with senior engineers who leave unempathetic comments and discuss how their feedback can be delivered constructively while still maintaining high standards. Emphasize that the goal of code reviews is to improve code quality and engineer growth, not to demonstrate superiority or catch mistakes. Many technical experts haven't been taught how to provide effective feedback, and a conversation about review techniques and tone can be transformative. Suggest specific approaches, such as starting with positive observations before making suggestions, or framing comments as questions rather than directives.

> This is a favorite interview technique that James uses with senior engineers. He asks them what they would do if approached by the junior engineer about what to do next after receiving a bad review. The answers tend to be deeply illuminating, especially when the answer is, "I'd tell them to suck it up because I learned the hard way too."

Systemically, this situation highlights the need for code review guidelines that establish expectations around tone, scope, and process. Some teams find it helpful to create templates that distinguish between must-fix issues and optional suggestions, or to establish norms like clearly defining appropriate language. James implemented a "code review buddies" system at a company, where junior engineers were paired with more experienced mentors who would review their code before the wider team reviewed it. This created a safer learning environment while still ensuring code quality. Over time, engineers naturally developed more constructive review habits, which fostered a better balance between thoroughness and empathy.

Making good technical decisions

Those lofty technical principles we introduced earlier also come into play when we start thinking about making good technical decisions. Your principles guide how you make the decisions that turn strategic initiatives or goals into reality. But translating your principles into everyday practice requires some governance. Typically, in a massive enterprise, this involves setting up structures like an architecture governance board, whose job is to oversee all architectural initiatives and ensure they align with established principles. We can hear you yawning already, and we agree: boards like this are essentially committees powered by presentation slides. They are where ideas go to die, or worse, where awful ideas are given terrible life—much as Dr. Frankenstein created his monster. Instead, we like lean and mean alternatives, of which our favorite is the RFC.

RFCs, or requests for comments, have been a common tool in engineering, especially when defining the standards for how the internet and internet-related protocols work. For example, RFC 822 is the original RFC that defines the structure of email messages. These are very formal documents that often take years to be reviewed and approved. However, many projects and companies have stolen the name and elements of their structure to create fast and practical architecture documents that can detail technical, operational, and even cultural changes they want to document.

So, what exactly does this new RFC species look like? Think of the modern RFC as your project's blueprint, but friendlier; it's a proposal that outlines significant changes or additions to your system. Imagine you're planning to introduce a shiny new feature, use a fresh new tool, or maybe tweak the fundamental architecture of your beloved app. Instead of diving straight into code and risking chaos, you pause and write an RFC. This doesn't just save you headaches; it actively invites your teammates into a thoughtful, structured discussion.

When you start using RFCs, you create a shared conversation space. Everyone can chime in: raising red flags, suggesting better solutions, or even offering a supportive thumbs-up. And here's the magic: not only does this reduce nasty surprises later, but it also genuinely deepens the team's shared knowledge. Have you ever worked somewhere where all the historical context lived only in someone's head? RFCs help avoid precisely that. By writing things down, you help everyone—current and future teammates—get up to speed effortlessly.

That shared conversation has guidelines, too: your technical principles. Our first step in reviewing any RFC is to validate that it aligns with our principles quickly. This is why the first example of a principle we posited is about simplicity. If you find yourself reading an RFC that seems like a maze of twisty passages, you're likely reading something too complex. Issues with other principles will also become readily apparent: modularity, maintainability, and scalability are all easy to determine when reviewing RFCs.

Many teams adopt RFCs because they want more explicit ownership and fewer of the surprise design-by-committee meetings. Almost everywhere that we've seen or introduced RFCs, we've seen an acceleration in the time it takes to make technical decisions. Before RFCs, decisions often dragged on without clear direction, ending in endless meetings and frustrating bikeshedding. After implementing a straightforward RFC process, the team could suddenly say, "This idea belongs to Ana," or "Jamal owns this decision." Clear ownership meant faster, bolder moves without the fear of failure looming. People felt freer to take smart risks because they knew the team had considered potential pitfalls.

In Practice: April's Story

Before April introduced RFCs in the studio operations team, technical decisions happened through "whoever builds it first, wins." This led to their current mess: three different logging systems because each senior engineer started building before discussing alternatives. April's first RFC requirement seemed simple: "Any new tool or pattern *requires an RFC." So, Marcus immediately tested this with an RFC proposing to rewrite the entire backend in Rust. But the RFC process worked exactly as intended: Sarah pointed out they'd need to hire three Rust developers (market rate: $300k each), and Derek calculated that the rewrite would take 18 months during which no features would ship. In addition, the juniors asked how this helped the company's immediate goal of shipping the 4K rendering feature. Marcus withdrew the RFC and submitted a new one: "Optimize hot paths in existing Node.js backend," which was approved and delivered a three-times performance improvement in three weeks.*

Here's how an RFC process typically looks: when you have an idea, you fill out a structured but simple template. Outline what you're proposing, why it's beneficial, possible drawbacks, and what alternative approaches you considered. Engage with the following example of an RFC template filled out to define the RFC for RFCs (yes, there are turtles all the way down) that helps explain how the workflow works in most organizations we've led.

Example: A Real-World RFC

This example shows an RFC about implementing the RFC process itself, which demonstrates the structure and approach while providing a meta-example of the process in action:

Title
RFC on RFCs

Main Author
@buritica

Backers
@kartar

Status
Draft, in review, rejected, approved

Summary
This proposal will structure how the [redacted] team makes technical decisions.

Motivation
By formalizing a technical decision process, we become a more effective and efficient team by enabling the following:

- Add a light, early stage of technical design that helps us understand our goals and evaluate decisions before writing heavy implementations.

- Create opportunities to raise risks or concerns based on our shared experience. One of us may have solved a similar problem before, but belong to a different team. This process enables us to share our expertise and past mistakes.

- Sharing institutional knowledge by exposing the entire team to different domain problems will improve the shared understanding of our platform.

- Centralize initial design documentation.

- Facilitate the future onboarding of new members, as they will be able to dive into the context of how we made certain decisions in the past.

- Update documentation by treating RFCs as live documents when past decisions need to be updated due to new information or features.

- Establish a shared sense of belonging by including our team in the process of important decisions.

- Diminished fear of failure by having a process where critical technical risks are identified. This will enable us to make bolder approaches as a team.

- Establish clear ownership. By enabling individuals and teams to propose and make critical decisions, we can take calculated risks, make mistakes, and enable continuous improvement.

- Encourage constant learning, by allowing members of diverse backgrounds to provide input.

- Gain clarity in the decision-making process by guaranteeing that those responsible for the decisions are leading the process, preventing bikeshedding, and reducing design-by-committee and brainstorming sessions.

Detailed Design: When to RFC

The distinction between when to create an RFC and when not to is not always clear. Engineers are expected to use good judgment when determining whether an RFC is necessary. Minor updates to endpoints may not require an RFC, but when in doubt, ask [redacted] whether specific work requires a proposal.

The following are some guidelines that may be used to determine whether a proposal is needed (though note that the cases do not always mandate a proposal):

- Will a new feature require a new endpoint?

- Are you starting a new project?

- Will your changes have an architectural impact?

- Are you proposing using a new tool, library, technology, or process?

RFCs may not be part of the initial process only. Sometimes, teammates may request a proposal to justify or argue modifications in an existing pull request—for example, when a new dependency is added to a system or a proposed modification has a broader system impact than initially expected.

Detailed Design: How to RFC

For new proposals, follow these steps:

1. Create an RFC template. One of the superpowers of RFCs is the consistency of their design, encouraging requests to consider different concepts with equal criteria.

2. Clone and fill that template in a new branch of a repository, create a Notion page, or manage your RFCs however you like.

   - We like prefixing a numbered 000X so we can track RFCs.
   - You may want to label an RFC in process with a WIP tag or title.

3. When ready for comments, open your pull request or share your RFC with your team.

4. Send a calendar invitation:

   - To: [redacted]
   - Title: [RFC] 000X deadline
   - All day event: Yes (marked as free)
   - Alert: on the day of the event
   - Description: link to RFC

5. All members of [redacted] are expected to read and add comments to the pull request or document within the expected timeframe. Members may indicate they've read the document and have no comments; votes are not required.

6. The author(s) and backers should resolve necessary comments and indicate which suggestions will not be addressed according to their judgment.

7. [Redacted] will approve by merging the pull request or marking the RFC as approved or vetoed, or may delegate someone for approval.

8. [Redacted] will generate a readable version and upload it to a shared document store so the entire company gets access.

For modifications to existing, previously approved proposals, follow these steps:

1. Clone and make modifications to the existing RFC.

2. Then follow the same process as for new proposals.

Additional Notes

Additional design notes include:

- If the confidence of acceptance is high (for architectural changes or new technology use), early work can be done while the RFC is under review, saving time.

- High-priority RFCs may have a review cycle of less than 24 hours, as long as they are brief and easy to understand. It'll be the author's responsibility to communicate the urgency of the proposal.

- The minimum time for an RFC to be reviewed should be enough to allow for time zone differences and reviewers' daily work and responsibilities. Use your judgment so teammates can review proposals without their own goals or deadlines being affected. Make sure you plan, and remember that you don't have to wait for approval to start work if you have high confidence in acceptance. When in doubt, check with [redacted].

- If comments or questions remain unresolved, the author may extend the time by modifying the date and the attached calendar invite.

Drawbacks

Adding RFCs as an initial step to our engineering process will increase the time it takes to ship something. This will improve as we ultimately incorporate RFCs into the way we build software.

Alternatives

Another option not detailed could be a forum-like platform, like Discourse.

What Is the Impact of Not Doing This?

Not having a straightforward way of making technical decisions will reduce our ability to grow as a team and to handle more complex problems under pressure. It will also limit the shared knowledge in our team and cause a loss of institutional knowledge by having implicit, undocumented processes and decisions.

Unresolved Questions

Some unresolved questions include:

- What parts of the proposal still need confirmation?
- How can we include folks outside of engineering in our RFC process?

Next steps

From the example, you can see that the format is straightforward but comprehensive. Then, once your RFC is ready for review, like many open source projects, it can become a pull request. In other organizations, they could be centralized in tools like Notion or internal wikis. Teammates then read it, drop comments, ask questions, or just mark that they've reviewed it. Then, after some healthy debate, the decision makers merge it in or approve it. And voilà—RFC becomes official documentation, clearly capturing why certain decisions were made.

RFCs are also living documents. They evolve as your project grows. If you've got new insights or changes that invalidate an old decision, then update the RFC! This fluidity makes RFCs incredibly valuable for maintaining an up-to-date narrative of your project's evolution. A good RFC process doesn't lock people out but invites them in. By sharing readable versions of RFCs, you can integrate nontechnical stakeholders. This keeps the whole organization in sync, not just the engineering crew.

What about architecture decision records?

You've probably also heard about architecture decision records or ADRs. We think they're worth mentioning as another form of technical architecture documentation. While RFCs help teams make decisions collaboratively before implementation, ADRs serve as the historical record of decisions already made. Think of RFCs as prospective decision-making tools and ADRs as retrospective decision documentation. Both are essential components of a comprehensive technical decision framework.

ADRs capture the context, decision, and consequences of architectural choices in a lightweight, standardized format. Unlike RFCs, which invite discussion and debate, ADRs document decisions that are final, providing future team members with the reasoning behind past choices. A typical ADR includes the problem context, considered options, the chosen solution, and expected consequences—both positive and negative.

The relationship between RFCs and ADRs creates a powerful decision lifecycle. An RFC might propose introducing microservices to solve scalability challenges, gather team input, and iterate through multiple revisions. Once approved and implemented, an ADR would document the final decision: "We chose to decompose our monolith into microservices to improve team autonomy and system scalability, accepting the trade-offs of increased operational complexity and network latency."

Many teams find value in automatically generating ADRs from approved RFCs, ensuring that decisions don't get lost in implementation details. Some organizations even reference specific ADRs in their RFCs, building on previous decisions and creating a connected web of architectural reasoning.

The key distinction: use RFCs when you need collaborative input on what to do, and ADRs to record what you actually did and why. Together, they create a comprehensive decision audit trail that prevents teams from repeatedly relitigating settled questions or forgetting the hard-won lessons from previous architectural choices.

> You can find some excellent ADR resources and examples in the GitHub repository (*https://oreil.ly/8DH5u*) by Joel Parker Henderson.

Leadership

Finally, we've discussed leaders' modeling behaviors frequently throughout the book: this applies to all aspects of communication, collaboration, and execution. And beyond modeling behaviors, leaders create the conditions for the success of all these aspects through their decisions on how they spend their time. When they dedicate time to code reviews, technical discussions, and knowledge sharing, they signal that these activities are valued. When they protect teams from constant context switching and unrealistic deadlines, they create space for thoughtful decision making rather than reactive firefighting. By aligning incentives with collaborative behaviors, recognizing team achievements, celebrating knowledge sharing, and rewarding cross-functional problem-solving, they build organizational cultures where quality software is built and shipped effectively.

Decision Making in Chaotic Conditions

All of these communication, collaboration, and execution steps require making a lot of decisions. In chaotic environments, the ability to make quick decisions is key. We want to streamline processes for reversible decisions, enabling teams to act swiftly without unnecessary debate. We also want to concentrate efforts on irreversible, higher-stake decisions and to spend time more effectively and reduce decision fatigue among team members. Effective decision making in chaotic conditions hinges on a blend of emotional intelligence, clear communication, and collaboration. To create the potential for this, we need to create an environment in which all of this can flourish.

Psychological Safety

Psychological safety is the belief that one won't be punished or humiliated for speaking up, and it is the foundation of effective collaborative decision making. Teams where members feel safe to share half-formed ideas, raise concerns, or admit mistakes make better decisions because they have access to more complete information and diverse perspectives. This safety isn't about avoiding critique or lowering standards; it's about creating an environment where the focus is on improving work rather than judging people. James worked with a team where the chief technology officer (CTO) had a habit of immediately shooting down ideas he disagreed with. Over time, engineers stopped proposing innovative solutions and instead defaulted to approaches that they knew the CTO would approve. The outcomes were a poorer quality of solution and a culture that was rife with low morale and lack of enthusiasm—and, in turn, high turnover and lack of interest in the technical outcomes.

In Practice: April's Story

April inherited a similar dynamic in the studio operations team, but it was peer driven. In architecture meetings, whoever spoke the loudest would win—usually Marcus. Sarah stopped proposing frontend improvements after Marcus called React Hooks "trendy garbage." Derek only spoke when directly asked for input, and the juniors didn't speak at all. So, April restructured these meetings: everyone wrote their proposals on sticky notes first (virtual or physical), and then discussed. Attempts to shut down discussion on bogus grounds were quashed by April. Suddenly, quiet Derek's suggestion to use database connection pooling saved $3,000 a month in AWS costs. Junior engineer Alex proposed a caching strategy that reduced API latency by 60%. Sarah felt safe enough to admit she didn't understand Derek's database indexes, leading to a knowledge-sharing session that helped everyone. Within a month, meeting participation went from three speakers to eight, and the quality of technical decisions improved dramatically.

Building psychological safety requires consistent, deliberate action from leaders. It means acknowledging your own mistakes openly, showing genuine curiosity about others' perspectives, and responding constructively to bad news. When you react to a production incident by asking "What can we learn?" rather than "Who screwed up?", you send a message about the team's values. Similarly, when you explicitly welcome dissenting opinions during decision-making processes and thank people for raising concerns, you create space for honest communication that is essential for effective decision making.

The benefits of psychological safety extend beyond better technical decisions to increased innovation, faster problem-solving, and higher retention. When team members feel safe taking interpersonal risks, they are more likely to share creative ideas, flag potential issues early, and invest in the team's success. They spend less energy on self-protection and more on solving problems collaboratively. Over time, this creates a virtuous cycle where trust begets more trust, and the team's ability to tackle complex challenges together grows stronger with each interaction.

Inclusive and Data-Driven Decision Making

Technical decision making is where principles and practices converge with the messy reality of competing priorities, tight deadlines, and limited information. The most effective technical decisions strike a balance between data-driven analysis and inclusive input from the team. While RFCs and other structured processes provide valuable frameworks, they're effective only when they genuinely include all relevant perspectives. Technical decisions often impact not just engineers but also product managers, designers, site reliability engineering (SRE) or operations teams, and other stakeholders. Inclusive decision making ensures these diverse viewpoints are considered, leading to more robust solutions and stronger alignment.

This can go too far as well. So, like all good things, do it in moderation. Think carefully about who to invite, and adjust based on feedback. We've all been invited to a meeting that we didn't need to be at. Good meeting etiquette, like having an "it's okay to vote with your feet and leave a meeting that you're not adding value to" is also useful here.

A common challenge is making technical discussions accessible to nontechnical participants. This doesn't mean simplifying complex topics, but instead focusing on outcomes and implications rather than implementation details when communicating across disciplines. Inclusive decision making also means considering diverse technical perspectives too. Junior engineers often hesitate to contribute to discussions dominated by senior voices, yet they can offer fresh perspectives unconstrained by "the way things have always been done." Different perspectives can uncover gaps in understanding and generate creative solutions that no individual would discover alone. This doesn't mean everyone gets an equal vote, but it does mean creating opportunities for all voices to be heard.

Transparency around decisions also strengthens inclusive decision making. When team members understand how and why decisions are made, they are more likely to accept outcomes, even when they disagree with them. Explicitly stating the factors being considered—such as technical merit, resource constraints, and business priorities—helps align expectations and reduce friction. It also helps team members grow and develop: if they want to progress professionally they need to understand how these components interact in making a team function.

Finally, use data. Whenever possible, back your decisions with data. Data helps eliminate personal biases and provides objective criteria for evaluating options. However, in chaotic environments, perfect data is rarely available. When you can't get comprehensive data up front, build proof-of-concept implementations or small experiments to gather empirical evidence. This approach lets you test hypotheses with minimal investment before making bigger commitments.

Continuous Improvement

Like all aspects of software development, collaborative decision processes benefit from regular reflection and refinement. Periodically reviewing the effectiveness of your RFC process, code review practices, or other collaborative mechanisms help identify opportunities for improvement and keep these processes aligned with your team's evolving needs.

Post-decision retrospectives can be particularly valuable. After implementing a significant technical decision, gather the team to discuss what went well in the decision process, what could have been improved, and what lessons should inform future decisions.

Continuous improvement isn't just about fixing problems; it's also about amplifying what works well. When teams identify effective practices, they can formalize and spread them throughout the organization. For example, a particularly successful approach to requirements gathering in one team might become a template shared across the company. Similarly, metrics that prove helpful in evaluating one type of decision might be adapted for other contexts. This positive reinforcement cycle helps build a culture of learning and adaptation, where teams constantly refine their approach based on real-world experience.

The most successful teams recognize that their decision processes shouldn't remain static as their context evolves. As teams grow, technologies change, and business priorities shift, the processes that served them well previously may need adjustment. Regular check-ins specifically focused on process health—separate from project-specific retrospectives—provide space to examine these broader patterns and make intentional changes. By treating their decision frameworks as products worthy of ongoing refinement rather than fixed rules, teams can maintain the right balance of structure and flexibility to navigate an ever-changing technical landscape.

Finally, as leaders, we play a crucial role in improving how the team makes good decisions. This doesn't mean making all decisions ourselves—quite the opposite. Not only is that counterproductive, but it's a fast path to burnout. Effective leaders create environments where good decisions emerge from well-structured processes and empowered teams, guiding without micromanagement. It's important to remember, though, that leaders are always still accountable for the decisions that are being made.

In Practice: April's Story

Three months into April's RFC process in the studio operations team, they ran their first retrospective on the process itself. The findings were illuminating: RFCs under two pages got quick approval, but anything longer stalled. Marcus admitted that he didn't read past page 3. Sarah pointed out that synchronous RFC reviews in meetings ate up too much time. The juniors hesitated to comment on senior engineers' RFCs, fearing they would look stupid. April iterated: RFCs now have a strict two-page limit with appendices for details. Reviews became asynchronous-first, with optional discussion meetings. They also added an "RFC buddy" system where juniors are explicitly assigned to review and must ask at least one question (removing the fear of whether they "should" comment). The refined process cut decision time from two weeks to three days while improving decision quality through more diverse input.

Leaders also model the collaborative behaviors they want to see in their teams. When leaders acknowledge uncertainty, consider diverse perspectives, and change their minds based on new evidence, they demonstrate the intellectual humility that enables effective team decision making. *Ted Lasso* popularized the phrase "Be curious, not judgmental," which is a tad reductive (and also not a quote by Walt Whitman). But it does highlight an approach we like: asking questions about potential decisions, rather than telling people the answers.

Conclusion: Collaboration Is the Glue

Collaborative technical practices and decision-making processes are the connective tissue that holds together your team, in the face of the chaos around them. It enables you and your team to achieve outcomes and reduce churn or flail. By establishing shared principles, consistent coding practices, structured decision frameworks like RFCs, and inclusive collaboration patterns, teams can navigate complexity and change while maintaining technical quality and velocity.

The most successful teams recognize that these practices aren't bureaucratic overhead but rather investments that pay dividends through reduced rework, faster alignment, preserved institutional knowledge, and more engaged team members. In chaotic environments where priorities shift frequently and information is often incomplete, strong collaborative practices provide stability and direction.

As you develop your collaborative technical practices, remember that the goal isn't perfect process adherence but merely somewhat better outcomes through thoughtful collaboration. Start with lightweight frameworks, adapt them to your specific context, and continuously refine them based on your team's experience. With patience and persistence, you'll build better software alongside a more resilient and capable engineering organization.

In Chapter 10, we'll talk about measurement, metrics, and how not to misuse them.

CHAPTER 10
Metrics That Matter for Engineering Teams

You're sitting in a meeting, and someone asks, "How's the engineering team doing?" There are a few things that happen next. Firstly, your stomach drops. Because explaining engineering productivity is complicated—especially to people whose understanding of the work of engineering is murky at best. Secondly, measuring that work is challenging. And thirdly, measuring it in a revelatory way is even harder. Suddenly, everyone is staring at dashboards full of squiggly lines going up and down, like a heart monitor during a horror movie. Some numbers are green, some are red, and nobody quite knows what any of it means for the actual health of your product or team.

If this scenario sounds familiar, you're not alone. The world of software engineering metrics can feel like navigating a maze with a broken compass. There are countless things you could measure, but figuring out what you *should* measure and how to interpret those measurements without losing your mind is tricky.

In this chapter, we will cut through the noise and discuss metrics that truly matter. These aren't vanity metrics that look impressive in a slide deck but tell you nothing useful, and they aren't metrics that make teams feel like an overeager hall monitor is watching them. Here, we will introduce you to the real and meaningful measurements that help you understand what's happening, identify problems before they escalate, and make informed decisions about where to focus your energy.

Why Measure Anything at All?

Before we dive into specific frameworks and numbers, let's address the elephant in the room: why bother measuring stuff in the first place? In chaotic environments, just keeping the lights on can feel like a victory. Adding yet another layer of tracking and reporting is a luxury you can't afford. But here's the thing: trying to improve without measuring means that you have no idea if you're improving. You might feel like you're making progress, but it's all based on vibes rather than reality.

Good metrics serve two essential functions:

1. They help identify problems before they become disasters. Trends in your metrics often reveal issues while they're still small enough to fix easily.

2. They indicate whether your improvement efforts are working. That fancy new CI/CD pipeline might feel like a win, but has it reduced your deploy times or failure rates?

That said, there's a dark side to metrics. When used poorly, they can create perverse incentives, demoralize teams, and lead to worse outcomes than having no metrics at all. Throughout this chapter, we'll discuss how to avoid these pitfalls and create a measurement system that helps, rather than hurts.

In Practice: April's Story

April discovered this truth during her first month at PixelCurl. When asked about team performance, the three senior engineers each told different stories. Marcus claimed that they were "crushing it" because the backend handled 10,000 requests per second. Sarah said they were struggling because feature delivery took forever. Derek insisted everything was fine because there hadn't been any database outages. Without shared metrics, they were three ignorant people describing different parts of an elephant. April implemented her first metric: feature cycle time from ticket creation to production. The results were shocking—average cycle time was 47 days, with features bouncing between the siloed engineers who couldn't review each other's code. This single metric transformed vague complaints into a concrete problem that they could address.

Metrics as a Product

Creating metrics for the sake of metrics is a common pitfall; it's often easy to collect data and even display it on a dashboard. Indeed, there is a classic scenario in observability: you spend 90% of the project building infrastructure and instrumentation, and then the remaining 10% is spent analyzing and visualizing the data. But data isn't insight. Your data has to solve a problem or provide answers. To ensure we focus on the insights and not just the data, try to think of your metrics as a product. They have customers, including you and other stakeholders, such as your leadership. And like any product, they don't need to be perfect at launch, but they should serve their users and get better over time. Treating them like a product helps you focus on what matters: solving real problems, delivering value, and setting clear goals.

While a whole book could be written on this topic, we'll focus on intentionally designing, balancing needs and wants, ensuring your metrics include human factors and psychological safety, identifying common pitfalls, and navigating evolution.

Design Intentionally

For metrics to be helpful, start with the basics: who they're for and why those folks need them. We do this by being very explicit about what we want: asking a question that is answered by one or more metrics and provides insight into the health of your world. Ensure your questions address real business needs; they shouldn't just be another layer of abstraction for collecting data without purpose.

Identify your customers and understand their needs. Your customers will likely fall into three groups: leaders outside of engineering, engineering leadership (including you), and individual contributors (such as engineers on your team).

You don't have to guess what your customers need: ask them. You've probably already gathered some data about friction points for your organization and team, but expand this to understand what insights each group needs by conducting some user research and identifying goals for each group:

Engineers
> For engineers, 95% of the time, it is metrics that relate to how they work and any friction inherent in their work. What metrics and insights would help them work more efficiently?

Engineering managers
> For engineering managers (and yourself), explore what metrics would help you (and your peers) identify bottlenecks and make improvements.

Leaders outside engineering
> For leaders outside of engineering, focus on their strategic concerns. Are they worried about costs, investment returns, or engineering velocity?

The Trap of Want Versus Need

There is a trap for new players in user research like this: sometimes what customers want and what they need are not the same. This trap is particularly prevalent in engineering metrics because stakeholders often ask for what sounds impressive or what they've seen elsewhere, rather than what would actually drive meaningful improvements. This trap is particularly endemic with regard to engineering velocity. James had a chief executive officer (CEO) who logged into GitHub and read through recent pull requests. In their next one-on-one meeting, the CEO declared that two engineers were far more productive than the others, while some others were labeled as lazy. When James asked how this CEO had worked that out, they indicated that the two standout engineers had committed thousands of lines of code, while others had committed hundreds of lines or fewer. James explained that the two most productive engineers were frontend engineers, who committed npm packages, JavaScript, and CSS (Cascading Style Sheets) code. He noted that the volume of code committed didn't reflect their productivity, but rather the nature of the code they worked on. The others, primarily backend engineers, committed far fewer lines of code. The CEO threw up their hands and asked, "So how do I know who is a good engineer or not?"

In Practice: April's Story

April faced similar pressure at PixelCurl. The vice president of product, seeing GitHub's contribution graph, declared Marcus the "MVP" because his graph was solid green while Derek's was mostly empty. April decided to pull actual data: Marcus committed 50 times daily (auto-formatting changes, console.log additions/removals, and constant WIP commits), while Derek committed once weekly with carefully crafted, fully tested features. April reframed the conversation: "What decision would you make differently if you knew who committed more often?" The vice president admitted that they wouldn't change anything. So, April shifted focus to deployment frequency and change failure rate—metrics that actually influence business decisions like release planning and customer communications.

Other questions we've both been asked include the following:

- Which team writes the best code?
- Which team is the most productive?
- How many hours do the engineers work every day?
- Which engineers work the most hours?

None of these questions is inherently bad, as they are attempts to quantify a domain with which the questioner is not comfortable. Naturally, we all ask questions of other professionals in our lives when we're trying to establish a baseline of understanding: accountants, plumbers, doctors, lawyers, and so on. So, don't demean someone for asking. Instead, provide them with better information. If you need to explain why a question or proposed metric is problematic or too simple—for example, a vanity metric like lines of code—then do so politely and suggest alternatives that will deliver real answers. It's your job to educate your leadership about the work you do and provide confidence in delivery, execution, and your leadership.

To do that, we must find the underlying question and answer it. So, for "Which team is the most productive," what are they really asking? It's likely the real question is about the predictability of delivery and potential bottlenecks in shipping software. Questions about hours worked and time tracking in general are likely related to confidence in delivery and the value provided. Ask more detailed questions to dig deeper, like the following:

- What decisions will these answers help you make?
- What would you do, or not do, differently if you had access to this information?

Reframe the question in the context of these answers. And in our opinion, in a significant number of cases, the questions actually boil down to, "Are we okay?" What people actually want to know is, "Will we meet our objectives? Are we getting value for money from our engineering investments? Is the quality of our software sufficient for our customers? Are we executing as we need to be?"

Constructing Questions

Before diving into solutions, review what you've learned. Take all the research and group it into key themes. You'll likely identify thematic groups with questions that apply to each customer but with differing levels of granularity. In answering the broader question of "Is our engineering velocity acceptable, and does it achieve our business goals?", for example, you can break this down into the following subquestions:

- Leadership
  — How fast are we shipping software?
- Engineering leadership
  — How many points of stories are we shipping?
  — How accurate are our estimates?
  — How many times do we deploy?
- Engineers
  — What is the cycle time from starting a story to reviewing and deploying the resulting pull request?
  — How long do CI or CI/CD runs take to build, test, and deploy code?

In many cases, the thematic groupings are ordered by granularity. For example, the "How fast are we shipping software?" query is a combination of the leadership and engineering questions that follow it. One of the easiest ways to ensure that you remain focused on getting the correct answers is to start with a big picture view and break it down.

All answers to our questions should have a metric and measurable success criterion; this helps validate the answers and makes it easier to associate them with a specific goal. Some questions might be more binary or have an endpoint—for example, achieving a goal and then entering a maintenance period. A good example here is CI tests, for which we aim to achieve a minimum coverage or reduce their runtime to below a certain threshold, and then monitor it.

In Table 10-1, let's break down one of these questions into component pieces.

Table 10-1. Example breakdown of a CI/CD-related engineering-velocity question into components, considerations, and measurable criteria

Component	Description
High level	Our engineering velocity is acceptable and meets our business goals.
Question	How long do CI or CI/CD runs take to build, test, and deploy code?
Success criteria	CI test runs take less than 15 minutes to run. CI/CD runs take less than 30 minutes to run and deploy.
Metrics	CI test run time CI/CD deployment run time
Maintenance goal	In 95% of both types of runs—CI and CI/CD—the success criteria are met or exceeded.

Choosing the Right Metrics for Your Context

By now, you might be thinking, "There are a lot of potential metrics. Do I need to track all of them?" And the answer is a resounding no. Metric overload is a real issue, and it's just as counterproductive as having no metrics at all. You don't need a stock of metrics; you need a focused set that provides clear insights into your team's health and performance.

The good news is that you don't have to reinvent the wheel. The DORA (DevOps Research and Assessment) metrics provide an excellent foundation because they're well-tested, focus on outcomes that matter, and can be implemented with most existing tooling. These four metrics capture the essential dimensions of software delivery performance. The DORA foundation consists of four metrics:

Deployment frequency
   How often you deploy to production

Lead time for changes
   Time from code committed to running in production

Change failure rate
   Percentage of deployments that require immediate fixes

Time to restore service
   How quickly you recover from incidents

These metrics work because they balance speed and quality while focusing on outcomes that your customers and business leaders care about. Most teams can implement basic versions using existing Git, CI/CD, and monitoring tools without significant additional investment.

Then, add context-specific metrics. This means one or two metrics based on your biggest current pain point that provide deeper insight.

If reliability is your biggest challenge, consider the following:

- Production incident volume by severity
- Mean time between failures (MTBF)

If delivery predictability is the issue, consider the following:

- Feature cycle time (idea to production)
- Sprint goal achievement rate

If quality concerns dominate, consider these:

- Bug escape rate (defects found in production versus preproduction)
- Testing coverage for critical paths

And if team health is suffering, consider the following:

- Developer satisfaction surveys (quarterly)
- On-call load distribution

This approach gives you five to six total metrics, which are enough to understand what's happening without overwhelming your team with data that they can't act on.

The implementation strategy includes the following steps:

Start small
    Implement the four DORA metrics first using existing tools.

Collect baseline data
    Gather four to six weeks' worth of data before making changes.

Add context
    Once DORA metrics are stable, add one or two context-specific metrics.

Review and adjust
    Quarterly, assess which metrics are driving decisions and which aren't.

Remember that the goal isn't to maximize any single metric but to maintain a healthy balance across all dimensions. Teams with the best long-term outcomes optimize for sustainable delivery of valuable, reliable software—something this focused set of metrics will help you achieve.

Take Stock of What You Already Measure

Before adding new metrics, understand what you're already tracking. Many teams are surprised to discover that they already have access to more data than they realize, particularly when implementing DORA metrics.

Common existing data sources include:

Git repositories
    Commit frequency, lead time data

CI/CD pipelines
    Deployment frequency, build success rates

Issue tracking systems
    Cycle times, incident tracking

Monitoring systems
    Uptime, error rates, recovery times

James worked with more than one team that was about to invest in an expensive new metrics platform but then realized they could extract most of what they needed from the systems they already had. By connecting their existing Jira, GitHub, and CircleCI data, they could track all four DORA metrics—plus additional context-specific measurements—without the need for additional tools.

Most modern development toolchains can support DORA metrics with minimal configuration. For example:

- Deployment frequency and lead time come from Git and CI/CD systems.
- Change failure rate can be tracked through incident management or monitoring alerts.
- Time to restore service often exists in your monitoring or ticketing systems.

Once you've inventoried your existing metrics, you'll face the classic build-versus-buy decision. Some teams discover that they can answer 80% of their measurement questions with a simple dashboard that pulls data from existing tools. Others find that a commercial platform's analytics and visualization capabilities justify the investment. The key is to start small; whether you build a lightweight integration or run a vendor proof-of-concept, validate that your chosen approach helps your team improve before

---

committing significant resources. The best metrics solution is the one your team will actually use.

Velocity: The Most Misused Metric in Software

Ah, velocity, the metric that everyone loves to hate. In theory, velocity is simply a measure of how much work a team can complete in a given time (usually a sprint). In practice, it's often weaponized in ways that make us want to hide under our desks.

Let's be clear: velocity is a capacity planning tool, not a productivity metric. Its primary purpose is to help teams predict how much work they can take on in future sprints. Using it to measure whether a team is "improving" or to compare different teams is about as useful as comparing the gas mileage of a sports car and a delivery truck.

Why Velocity Goes Wrong

The most common velocity mistakes we see are as follows:

Treating it as a performance metric
"Team A has a velocity of 50 points while Team B only has 30. Clearly, Team A is better!" No, no, no. Points are arbitrary and unique to each team. Perhaps Team B has more challenging problems, greater technical debt, or simply sizes their stories differently.

Setting targets based on past peaks
"You reached 45 points once, so that's your new baseline!" This encourages teams to game the system by inflating estimates or taking shortcuts that create technical debt.

Comparing velocity among teams
Different teams estimate differently, work on different types of problems, and have different team compositions. Comparing their velocities is like comparing apples to spaceships.

James tells a painful story about an organization that decided to "standardize" story points across teams to make velocities comparable. The result? Teams argued more about estimation than actual technical problems. Some teams artificially inflated their points to appear competitive, while others became demoralized when their "velocity" seemed lower than others. Eventually, management abandoned the approach after realizing they had created a competition that was actively harming product quality.

How to Use Velocity Correctly

If you do track velocity (and it can be helpful when used right), here's how to keep it healthy:

Use it for capacity planning.
You can make considerations like, "Based on our average velocity, we can probably take on these five stories next sprint."

Look for dramatic changes.
A sudden drop or spike in velocity might indicate a problem worth investigating.

Keep it within the team.
Velocity is most useful as an internal team metric, not something to report to upper management.

Remember that the goal isn't to maximize velocity; it's to deliver valuable, high-quality software at a sustainable pace.

James worked with a team that tracked their velocity privately, within the team. They used a rolling three-sprint average to plan upcoming work and kept an eye out for unexpected changes. When their velocity dropped by 40% over two sprints, they had an honest conversation about what was happening and discovered that technical debt in a critical component was slowing them down. Because this metric wasn't judging them, they could safely bring the issue to their product owner and make the case for spending a sprint addressing the technical debt.

Build Your Metrics with Your Customers in Mind

Your solutions will take different shapes for different audiences. While dashboards are common, they're not the only way to communicate. Your metrics toolkit might include reports as source material for workshops or brainstorming sessions, and these reports can be attached to runbooks and activities. Observe how your audience consumes and interprets the metrics, and be prepared to adapt accordingly. If you notice teams struggling to use your dashboards or make sense of the metrics, conduct further user research to determine if the metrics need to be adjusted or if they are not suitable for their intended purpose.

The Human Side of Metrics

Let's not forget the human element. Metrics exist to serve people, not the other way around. We briefly discussed this when we explored the want-versus-need trap, but it's also a broader concern: if you *want* to drive improvement rather than fear, you *need* to create psychological safety around metrics. There are some simple steps you can take to ensure your metrics are safe:

Measure teams, not individuals.
Individual metrics almost always create unhealthy competition and behaviors. However, this doesn't mean managers should ignore individual performance entirely. The key distinction is between *measuring* individuals and *identifying* those who need support.

How to Identify Underperforming Team Members Without Individual Metrics

Individual performance issues typically surface through qualitative signals before they appear in quantitative data:

Code review patterns
Are their pull requests consistently requiring extensive rework? Are they struggling to get reviews approved?

Collaboration signals
Are they participating in team discussions? Are they asking questions when they're stuck, or helping their teammates?

Delivery patterns
Are they consistently missing sprint commitments? Are they taking much longer on similar-sized tasks than their peers?

Knowledge sharing
Can they explain their work clearly? Are they learning from feedback?

Use metrics as context, not judgment.

When you notice these qualitative concerns, team-level metrics can provide helpful context. If the team's cycle time has increased and you've observed one person struggling with code reviews, that's a coaching opportunity, not a performance evaluation.

Focus on support, not surveillance.

The goal is identifying when someone needs help—whether that's mentoring, training, creating clearer requirements, or addressing external factors affecting their work. Frame conversations around "How can I help you be more effective?" rather than "Your numbers are low."

Involve teams in choosing metrics.

People support what they help create. Teams should have input on which metrics are most relevant to their work. When teams participate in defining their measurements, they're more likely to use them constructively.

Create a blameless culture.

When things go wrong (and they will), focus on learning rather than placing blame. This applies both to system failures and individual performance issues. A team member struggling with delivery might be dealing with unclear requirements, technical debt, or knowledge gaps—all systemic issues that metrics alone won't reveal.

In Practice: April's Story

April learned this lesson painfully. She initially tracked pull request review time by reviewer, hoping to identify bottlenecks. The data showed that Derek took three-times longer than others to review. When she shared the metric publicly, Derek started approving reviews without real scrutiny to improve his "score." Bug escape rates tripled in the next two weeks. April dug deeper: Derek's reviews caught 89% of production bugs before shipping, while faster reviewers missed critical issues. April immediately changed approach: she tracked the team-level review time and bug escape rate together, showing that thorough reviews correlate with fewer production issues. She celebrated Derek's thoroughness while pairing him with juniors to share his bug-spotting expertise.

The best individual performance insights come from regular one-on-ones, peer feedback, and direct observation of work quality and collaboration patterns. Metrics should inform these conversations, not replace them.

Balancing Quantitative and Qualitative Data

We also recommend striking a balance between the use of quantitative and qualitative data. Hard numbers tell part of the story, but not all of it, and the why behind a metric's value might not be anything technical. To capture some of this information, you're probably already doing some of what you need. If you're conducting retrospectives on sprints, include questions about morale and how people are feeling, and ask participants to share concerns, friction points, and issues. You've also got clients that you deliver products and services to, so ask them about their experiences. Frameworks like NPS (net promoter scores) are straightforward to implement. Even easier is the happy/neutral/sad approach—for example, post-ticket surveys where customers click on a happy face, neutral face, or sad face. You can also embed this into product features and directly ask customers for feedback using a single one-click tool like this. The richest insights often come from combining quantitative trends with qualitative explanations.

Humans and Metrics

It's also important to remember that the people on your team will not only interpret metrics but they'll also change their behavior in response to them.

Goodhart's Law

Goodhart's Law states that "When a measure becomes a target, it ceases to be a good measure." Once people know they're being measured, they will optimize for that metric—sometimes at the expense of what actually matters. For example, if teams measure only feature development or availability, they will tend to focus solely on building features or prioritizing availability at the expense of performance. This also applies to nontechnical domains, such as costs (as discussed in Chapter 7). Focusing on cost savings sometimes costs you more money in the long term because the quality of service suffers. A good way to avoid getting bitten here is to focus on outcomes as well as volumes—for example, recognizing that it's great that you built all those features, but also assessing whether they were good features that met your customers' needs.

In Practice: April's Story

April experienced Goodhart's Law firsthand when PixelCurl's board demanded "improved velocity." She increased story points from 30 to 45 per sprint. Was it a success? Not quite. The team achieved this by marking stories complete without documentation, skipping code reviews for "trivial" changes (which weren't in fact trivial), and deferring all technical debt work. After two months of "improved velocity," the system became so fragile that a routine deployment took down production for six hours. April presented the correlation to leadership: when velocity became the target, quality metrics (test coverage, code review thoroughness, documentation completeness) all declined. She implemented a balanced scorecard: velocity, quality, and stability metrics must all stay within acceptable ranges.

Also consider the concept of demand characteristics (*https:// oreil.ly/FkOzg*), which is seen in experimental studies, where participants form an interpretation of the experiment's purpose and subconsciously change their behavior to fit that interpretation. Or in our case, people make assumptions about why something is being measured and subconsciously behave differently.

Context ignorance

Metrics without context are meaningless at best, and misleading at worst. We saw this in the discussion about want versus need. Never share metrics without some color and context that explains the following:

- What the metrics measure
- What the metric value means, especially if it looks like an outlier or an oddity

You only have yourself to blame if you provide an ambiguous metric that causes the audience to make assumptions or having to fill in the gaps themselves—because they'll always assume something that you don't expect, or don't want.

"What gets measured gets done"

This sounds awesome, right? We measure things, and that provides a focus for the team, enabling them to get things done. There is a downside, though: anything not being measured tends to be overlooked. As we've said, you can't measure everything, so you're always making trade-offs about what deserves attention and what doesn't.

The problem of invisible work is real. Think about all the crucial activities that keep your systems running but rarely show up on dashboards—the time spent mentoring junior engineers, the hours of documentation writing, the careful code reviews that prevent future disasters, or the relationship building with stakeholders that smooths the path for future projects. These activities aren't typically measured, yet they're fundamental to long-term success.

This creates a particularly nasty trap for teams trying to "do the right thing." If your metrics focus exclusively on velocity and feature delivery, what disappears? Technical debt remediation, security improvements, and infrastructure upgrades all take a back seat because they don't move the needle on what's being measured. Teams know these things matter, but when performance reviews and team goals are tied to specific metrics, the unmeasured work becomes career limiting.

The solution isn't to measure everything—in that, lies chaos and metric fatigue. Instead, acknowledge the limitation explicitly. When rolling out metrics, be clear about what you're not measuring and why that work still matters. Consider rotating your focus metrics quarterly or semiannually to ensure that different aspects of the work receive their time in the spotlight.

Most importantly, create space for the unmeasured work. Google's famous "20% time" (*https://oreil.ly/7zXvQ*) wasn't just about innovation—it was an acknowledgment that not everything valuable can or should be measured. Whether it's dedicated time for technical debt, learning, or relationship building, protecting time for unmeasured activities ensures they don't get completely crowded out by the tyranny of the dashboard.

Metrics are a tool for understanding and improving your work, not a complete definition of what constitutes valuable work. The best teams understand this distinction and act accordingly.

Make Metrics Visible and Accessible

Metrics hidden in a dashboard that nobody looks at might as well not exist. Make your key metrics visible:

- Place dashboards on team TVs or shared screens.
- Ensure automated reports are sent to team communication channels.
- And don't forget the reviews we mentioned earlier in team meetings and beyond.

The goal is to make metrics an integral part of everyday conversation, not something reserved only for quarterly reviews.

Metrics Need Ongoing Maintenance, Just Like Code

Track versions and changes you make to metrics. Keep a changelog of what changed and why. When you update how a metric is calculated, document it. Your future self (and everyone else) will thank you. For example, if you change how you measure pull request review time to exclude weekends, that's a big deal—write it down! (Or better yet, create a new metric and give it a meaningful name so it's easy to understand.) This helps teams understand why their numbers might look different and maintains trust in your metrics.

Think of versioning in this way: small tweaks receive a minor version bump (e.g., v1.1 to v1.2), while more significant changes that may impact historical comparisons warrant a major version update (e.g., v1.0 to v2.0). This makes it easier to track changes over time and understand when metrics might not be directly comparable.

Make ownership crystal clear. Someone needs to own these metrics—both the technical implementation and the business value they provide. Without clear ownership, metrics tend to get boring or break without anyone noticing. Your metrics owner should do the following:

- Keep an eye on data quality and reliability.
- Check that metrics still align with team goals.
- Handle questions and concerns from teams.
- Plan and prioritize improvements.

This doesn't mean one person does everything, but someone does need to be responsible for the health of your metrics system. Think of them as the product manager for your metrics.

Listen to your users and evolve. Your metrics are a product, remember? That means gathering feedback from your users (the leaders, managers, and engineers we discussed earlier) and using it to drive improvement. Some ways to do this include the following:

- Regular check-ins with teams using the metrics
- Surveys about metric usefulness and pain points
- Watching which metrics actually drive decisions (and which ones nobody uses)
- Looking for gaps where teams need data but don't have it

Sometimes you'll find metrics that seem great in theory but aren't actually helping anyone make decisions. That's fine; eradicate them! It's better to have fewer, more valuable metrics than a dashboard full of numbers that nobody uses.

Stay ahead of changes to your products, platforms, and environments. Your environment is constantly evolving, with new tools, updated platforms, and different processes. Your metrics need to keep up. Keep an eye on the following:

New data sources becoming available
    This means assessing whether that shiny new CI system has better build metrics.

Existing tools being deprecated
    This is the reverse, where consideration is needed as to whether you have a plan for when the old build system is removed.

Changes in how teams work
    If everyone switches to trunk-based development, your pull request metrics might need updates.

Remember to phase out your metrics. Not every metric needs to live forever. When a metric stops being useful, remove it. You may have solved the problem it was tracking, or teams may have moved on to different challenges. Having too many metrics can be just as detrimental as having too few, as it makes it harder for teams to focus on what truly matters.

Before removing a metric, check on the following:

- If anyone is still using it for decision making
- If it's part of any regular reports or dashboards
- Ensuring you've stored historical data for compliance or comparison
- Considering whether there's a newer metric that better serves the same purpose

So, what is the bottom line? Your metrics aren't a set-it-and-forget-it thing. They need regular attention to stay relevant and reliable. However, with clear ownership, effective feedback loops, and a willingness to evolve, they'll continue to help teams make better decisions.

AI Doesn't Change Your Approach to Metrics

As AI tools become integrated into development workflows, your fundamental metrics approach remains sound; you still need balanced metrics that drive decision making and improvement. What evolves isn't the methodology but the specific measurements and how you present them. Your metrics product should adapt by creating segmented views that compare AI and non-AI performance, highlighting both efficiency gains and quality impacts. This approach incorporates feedback loops to identify where AI tools deliver the most value and introduces new visualizations that track the impact of AI on collaboration patterns.

For productivity measurement, the key dimensions remain equally valuable for AI-augmented development. You'll want to measure developer satisfaction with AI tools and their impact on work quality, while tracking concrete outcomes (such as defect rates or feature completion times) with AI assistance. Monitor AI usage patterns, generated code volume, and pull request throughput to understand activity levels. Pay attention to how AI affects collaboration between team members and across teams, or how it helps create or replace documentation. Most importantly, measure the time saved and workflow changes resulting from AI adoption to quantify efficiency gains.

Beyond traditional productivity metrics, you may need to add new dimensions, particularly around trust. Measuring engineers' trust in AI-generated code, comments, and recommendations becomes crucial because this trust dimension directly affects adoption patterns and the ultimate value that teams derive from AI tools.

Your metrics strategy should anticipate the questions that stakeholders will inevitably ask about AI adoption. Where is AI providing the most value? Are we seeing quality trade-offs? Which teams are benefiting most from these tools? Having these insights ready builds confidence in your AI implementation strategy while helping teams make better decisions about where and how to leverage these tools. The teams that succeed with AI are those that measure its impact thoughtfully, adjusting their approach based on data rather than assumptions about how AI should theoretically improve productivity.

Conclusion: Measurement as Illumination

Throughout this chapter, we've explored how to create metrics that help rather than hinder your engineering teams. The key insight running through every example and framework is that metrics should serve as guides on your journey rather than become the destination itself.

Effective metrics illuminate what's happening in your engineering organization, helping teams make informed decisions about where to focus their energy. They reveal hidden bottlenecks, validate or challenge your assumptions about productivity, and help you prioritize improvements that matter. But remember, they're instruments for understanding, not hammers for judgment. When metrics transform from measurement tools into performance targets disconnected from real outcomes, they lose their ability to drive genuine improvement and often create harmful behaviors instead.

As you develop your metrics strategy, keep these principles in mind. Focus on measuring outcomes that matter to your customers and business, not just outputs that clutter your dashboards. Recognize that speed, quality, cost, and team well-being exist in constant tension; optimizing for any single dimension while ignoring the others leads to predictable failures. Treat your metrics program as you would any other product by understanding your users' needs, gathering feedback regularly, and iterating based on what you learn. Start simply with a handful of carefully chosen measurements rather than overwhelming teams with dozens of confusing datapoints. Most importantly, build psychological safety around your metrics so that teams view them as tools for learning and improvement, rather than as weapons for blame.

The real value of metrics lies not in the numbers themselves but in what they enable. Good metrics spark meaningful conversations about improvement, reveal insights that challenge conventional thinking, and inspire teams to experiment with better ways of working. When used thoughtfully, they become powerful allies in your quest to build not just better software, but better teams and more sustainable engineering cultures. The goal isn't to become beholden to the dashboard but to use data wisely in service of creating environments where both people and products can thrive.

In the next (and final!) chapter, we'll summarize everything you've learned so far.

CHAPTER 11

Fitting It All Together

If you've made it this far, you already know: chaos doesn't give out medals for endurance. But you're still here, and that matters. Perhaps you're reading this at your desk between incidents, on your phone during an excruciating planning meeting, or at home after another day of trying to bring order to chaos. Wherever you are, take a moment to acknowledge that you're still here, still fighting, still trying to get better at this impossible job.

That persistence matters more than you might think. Because engineering leadership in chaos isn't about having all the answers or following a perfect playbook. It's about showing up, day after day, and making things a little bit better than they were yesterday. It's about building momentum when everything wants to grind to a halt. It's about creating clarity when fog is the natural state. And most importantly, it's about helping your team thrive despite conditions that would make most people throw their hands up and walk away.

We'd also be remiss not to remind you that sometimes, walking away is the right choice. Persistence is powerful, but not at the expense of your mental and physical health. Sometimes situations can't be improved or fixed. If you feel burnt out or that you're treading water without making forward progress, think carefully about what the best outcome is for you. Taking care of yourself first is always the right choice.

Throughout this book, we've thrown a lot at you: technical principles, user stories, budget spreadsheets, vendor management, requests for comments (RFCs), retrospectives, metrics that matter, and metrics that can mislead. Each chapter has its own universe of chaos and disorder, along with approaches, tools, and techniques to deal with that. Now it's time to step back and see how it all connects, because the real power isn't in excelling in any single aspect of engineering leadership; it's in understanding how these pieces work together to create something greater than the sum of their parts.

The Symphony of Chaos

As you've discovered in this book, we both love a metaphor. So here's (almost) your final one: the reality of engineering leadership.

As the engineering leader, you're the conductor of the symphony—the maestro—and it's up to you to guide your team—strings, brass, woodwinds, percussion—through the chaos with the cues you give from your podium and the feedback you receive from your orchestra as they perform. You stand at the center of a complex system, each part demanding your attention. Your people form the foundation, the core strength of everything you build. Your technical infrastructure is essential and enabling, yet requiring constant vigilance. Your processes flow beneath the surface, often invisible but critical to keeping work moving forward. Your metrics provide the necessary feedback, confirming whether what you're building can withstand pressure. And you're there in the middle of it all, armed with incomplete information and an evolving plan, working to align everyone toward the same goal while priorities shift, requirements change, and unexpected crises demand immediate response.

In calm conditions, you could focus methodically on each element. You'd have time to be deliberate, to refine your approach, and to build systematically toward excellence. In the reality of constant change, you're managing shifting timelines, team transitions, and strategies that evolve weekly, all while navigating urgent demands from every direction. Success isn't about creating the perfect system under ideal circumstances; it's about maintaining forward momentum and delivering value despite the turbulence.

This is why everything we've discussed in this book interconnects so deeply. Your technical principles aren't just about the quality of your code; they shape how your team collaborates and makes decisions. Communication isn't just about sharing information; it builds the psychological safety that enables innovation. And your metrics aren't just numbers on a dashboard; they are the feedback loops that tell you whether your other efforts are working. Every aspect of engineering leadership influences every other aspect, creating either virtuous cycles that build momentum or vicious cycles that compound dysfunction.

The Foundation: People and Safety

Let's start with the most important aspect: people. We've said it before, but it bears repeating because it's so easy to forget when you're drowning in technical debt and deadlines: without psychological safety, nothing else works. If there is no trust, there is no team. Instead, it's just a group of people cashing checks and keeping their heads down. It's not your beautiful technical architecture, your elegant processes, or your comprehensive metrics—it's nothing.

When your team doesn't feel safe to speak up, admit mistakes, challenge ideas, or show vulnerability, you're operating at a massive disadvantage. You might not see it immediately. People are good at hiding their fears, going through the motions, and delivering just enough to avoid attention. However, over time, the absence of psychological safety creates a kind of organizational scar tissue that makes everything more difficult. Innovation slows down because people are unwilling to take risks. Problems fester because no one wants to be the messenger. Technical debt accumulates because admitting its existence feels like admitting failure.

Building psychological safety in chaotic environments is particularly challenging because chaos itself often feels unsafe. When priorities shift daily, when leadership changes frequently, or when the company's future feels uncertain, people naturally become defensive. People protect themselves by keeping their heads down, avoiding anything that might make them a target, and doing precisely what they're told—and nothing more.

This is where your role as a leader becomes crucial. You become the stability, not by pretending the chaos doesn't exist or by promising you can stop it, but by being consistent in how you show up, how you respond to problems, and how you treat people. When someone raises a concern, you listen. When someone makes a mistake, you focus on learning. When someone challenges your ideas, you thank them for their candor. These small acts, repeated consistently over time, create pockets of safety that gradually expand until they encompass the entire team.

But safety alone isn't enough. You also need capability and cohesion. A safe team that can't execute is just a support group. This environment is where all those processes and practices we've discussed come into play. They become the scaffolding that enables teams to work together effectively. User stories create shared understanding. Code reviews spread knowledge. RFCs encourage thoughtful design. Retrospectives promote continuous improvement. Each practice, when implemented thoughtfully and maintained consistently, adds another layer of capability to your team.

The Lighthouse: Direction, Not Drift

Once you have a foundation of safety and capability, you need direction. This might seem paradoxical in chaos, where the future is inherently unpredictable. How can you set a direction when you don't know what next month will bring, let alone what next year will bring? The answer is that direction in chaos isn't about predicting the future; it's about creating coherence in the present.

Think of direction like this: you're not plotting a course to a specific destination, you're choosing which general direction to head. North might be "improve reliability." Northeast might be "reduce technical debt while expanding into new markets." East might be "rapid feature development." You might not know exactly where you'll end up, but you know which way you're moving—and, more importantly, which ways you're not moving.

This is where those technical principles we discussed become so powerful. They're not rigid rules to constrain your team; they're guideposts that help your team make consistent decisions even when specific requirements are unclear. When an engineer faces a choice between a simple solution and a complex one, "simplicity first" provides direction. When a team debates whether to document a decision, "documentation as communication" settles the question. These principles create alignment without requiring constant coordination, which is essential when you don't have time (or the desire) for endless meetings and debates.

Setting direction also means making peace with incompleteness. In chaos, you'll never have all the information you want, you'll never be certain that you're making the right choice, and you'll never be able to predict all the consequences of your decisions. But as we've emphasized throughout this book, action beats inaction almost every time. A decent decision made today is usually better than a perfect decision made next month. Because in chaos, next month's context will be completely different anyway.

This means being thoughtfully decisive by gathering sufficient information to make a mostly reasonable choice, communicating that choice, and then committing to it long enough to see results. It means being willing to adjust when new information emerges, but not second guessing yourself every time someone raises a concern. It's a delicate balance, and you won't always get it right. But the teams that thrive in chaos are those whose leaders can make decisions and stick with them long enough for the team to build momentum.

The Engine: Process and Execution

With safety, capability, and direction in place, you need an engine that turns potential into reality. This is where process comes in. But this isn't process as most people understand it—not a heavy, rigid, bureaucratic process that slows everything down. Instead, we're talking about a lightweight, adaptable process that enables execution rather than constraining it.

The key insight about process in chaos is that it should be just enough to prevent disorder, but not so much that it prevents adaptation. Think of it like the minimal amount of structure needed to keep things from falling apart: a daily standup that actually stays under 15 minutes and helps people coordinate, a simple Kanban board that makes work visible without requiring hours of grooming, or code review that catches problems without becoming a bottleneck. Each process should earn its keep by clearly making things better, and the moment it stops adding value, it should be simplified or eliminated.

This is why, throughout the book, we've emphasized starting small and building up. It's tempting, especially if you're coming from a more structured environment, to try to implement all the processes you're familiar with—sprint planning, backlog grooming, story pointing, retrospectives, demos, architecture reviews, and on and on. But in chaos, every process has a cost in terms of time, energy, and cognitive load. Your team only has so much capacity for process, and if you use it all up on ceremonies and meetings, there's nothing left for actually building things.

The art is in finding the minimal viable process for your current situation, and then evolving it as conditions change. When you're in pure survival mode, all you need is a daily check-in and a shared list of priorities. As things stabilize, you should add lightweight planning sessions. As the team grows, you need more structured communication patterns. The process should grow organically in response to actual problems, not in anticipation of theoretical ones.

Execution in chaos also requires a different mindset about what constitutes success. In stable environments, you can aim for perfection, or at least something close to it. You can take the time to get things exactly right—to polish every edge, to optimize every algorithm. In chaos, perfection is the enemy of progress. You need to embrace "good enough" as a legitimate target, at least initially. In chaos, polished work is often irrelevant work. Ship the feature that solves 80% of the problem, and then iterate. Build the system that handles the common cases, and then add edge case handling as needed. Write the documentation that helps people get started, and then expand it based on actual questions.

It's not about lowering standards; it's about recognizing that in chaos, momentum matters more than perfection. A team that ships something imperfect but useful every week will outperform a team that spends months crafting the perfect solution. Because by the time that ideal solution is ready, the problem it was supposed to solve has probably changed. The key is to be intentional about what you're trading off and to have a plan for addressing those trade-offs over time.

The Compass: Metrics and Measurement

As your engine runs, you need a way to know if you're actually making progress. This is where metrics come in—but again, not in the traditional sense. Not vanity metrics that make you look good in presentations. Not weaponized metrics used to punish teams that don't meet arbitrary targets. Instead, we're talking about metrics as a compass, helping you understand where you are and whether you're moving in the right direction.

The biggest mistake is trying to measure everything, with dozens of metrics and endless dashboards. In chaos, you don't have time to analyze all that data, and most of it isn't actionable anyway. You end up with information overload that obscures rather than illuminates what's really happening.

Instead, focus on a small number of metrics that directly relate to your current challenges. If reliability is your biggest problem, track mean time to recovery and change failure rate. If delivery is the issue, measure cycle time and deployment frequency. If quality is suffering, monitor bug rates and test coverage. The key is to be selective and purposeful, and to choose metrics that will actually influence decisions rather than just creating pretty charts.

Remember, too, that metrics in chaos need to be treated differently from those in stability. In stable environments, you can set targets and expect steady progress toward them. However, in chaos, metrics will be messy, with wild swings that may have nothing to do with your team's performance. A deployment might fail because a vendor had an outage. Cycle time might spike because half the team got pulled into an emergency. Bug rates might jump because you finally started tracking them properly.

This is why context matters so much with metrics. Raw numbers without explanation are useless, and they're actively misleading. Every metric should have a story that explains what it means, why it might be changing, and what (if anything) should be done about it. This requires more work than just making dashboards, but it's essential for maintaining trust and making good decisions. Metrics without context are just numbers cosplaying as truth.

The Ecosystem: Building Beyond Your Team

Your team doesn't exist in isolation, especially during times of chaos. You and your team are part of a larger organism that includes other engineering teams, product, design, sales, support, leadership, and customers. Success requires navigating effectively, building bridges where they're needed, and sometimes creating boundaries where they're not.

This is where all that work on communication and collaboration pays off. The RFC process isn't just about making better technical decisions; it's about creating visibility into those decisions for other teams. The documentation you write isn't just for your future self; it's for the support engineer trying to help a customer at 2:00 a.m. The metrics you track aren't just for your team's improvement; they're for building confidence within leadership that the engineering team is delivering value despite the chaos.

Effective management in general becomes particularly crucial in chaotic environments. Your leadership teams are probably feeling the pressure too, dealing with their own chaos at the organizational level. So, they need confidence that engineering is under control, that problems are being addressed, and that progress is being made. This doesn't mean hiding problems or pretending everything is fine. Rather, you need to be proactively communicating what's happening, what you're doing about it, and what you need from organizational leadership.

The vendor and budget management aspects we discussed aren't just about controlling costs; they're about building an extended capability network. In chaos, you can't build everything yourself. You need partners who can provide specialized capabilities, handle noncore functions, and scale with your needs. But these relationships require active management. Vendors need to understand your constraints and priorities. Contracts need to be flexible enough to adapt to changing conditions. Costs need to be predictable enough to plan around, even when everything else is unpredictable.

Cross-functional collaboration becomes even more critical in chaos because misalignment is incredibly expensive. When engineering teams build the wrong thing because requirements were unclear, when the product team promises features that can't be delivered, or when sales teams set expectations that can't be met, the resulting friction can bring everything to a halt. This is why investing in collaborative practices, shared language, and mutual understanding pays such high dividends. Every hour spent aligning with the product team saves days of rework. Every conversation with sales teams prevents a crisis with a customer. Every sync with the support team reveals problems that you didn't know existed.

The Evolution: From Chaos to Capability

Here's something we haven't talked about explicitly but that underlies everything in this book: chaos doesn't last forever. Organizations either evolve beyond chaos, find stability and scale, or they fail. Your job as an engineering leader isn't just to survive the chaos; it's to help your organization evolve through it.

This evolution doesn't happen overnight, and it doesn't happen linearly. You'll have periods of progress followed by setbacks. You'll solve one source of chaos only to discover another. You'll build capabilities that become obsolete as the organization changes. This is normal and expected. You just need to focus on maintaining forward momentum, even when it feels like you're taking two steps forward and one step back.

Each of the practices and approaches we've discussed in this book builds organizational capability that persists beyond the immediate chaos. The psychological safety created becomes part of the culture. The technical principles that you establish will guide future decisions. The processes that you implement will evolve. The metrics that you establish will create feedback loops that drive continuous improvement.

Layer by layer, you're building an organization that can handle not only the current chaos but also future challenges.

This is why we emphasize starting small and iterating. Instead of trying to transform everything at once, you're planting seeds that will hopefully grow over time. The RFC process might evolve into a sophisticated architecture review system. Basic metrics might become the foundation of an engineering insights platform. That daily standup might transform into a highly coordinated delivery system. However, none of this happens overnight, and attempting to skip directly to the end state usually fails.

The evolution also applies to you as a leader. Every challenge you navigate, every crisis you resolve, and every team you build adds to your capability. You develop instincts about what works and what doesn't. You learn to read the signs of impending problems. You build a network of relationships that can provide support and guidance. You create a reputation that gives you credibility and influence. These accumulated capabilities make you more effective over time, able to handle greater challenges with less stress.

The Reality: It's Harder Than It Looks

We need to be honest about something: everything we've discussed in this book is more complicated to implement than it sounds. When we say "build psychological safety," it sounds straightforward. But in reality, it requires hundreds of small actions, consistent behavior over months or years, and the ability to maintain that consistency even when you're exhausted, frustrated, or dealing with your own challenges.

When we say "set clear direction," it may seem simple. but in reality, you're making decisions with incomplete information, and then communicating those decisions to people who may disagree, while maintaining conviction even when you're not entirely sure yourself. It means being wrong sometimes (or a lot) and admitting it without losing credibility. It means changing course when necessary without appearing indecisive.

When we say "implement lightweight processes," it might sound easy. However, it actually means constantly fighting against process inflation, pushing back against well-meaning additions that would slow things down, and maintaining discipline when everyone around you is adding complexity. It means being the one who says no to the process that would make one person's work easier, because it would make everyone else's work harder.

When we say "track meaningful metrics," it might sound obvious. But, in reality, it means resisting the pressure to measure what's easy rather than what matters, explaining again and again why certain metrics are misleading, and maintaining faith in your measurements even when they show things getting worse before they get better.

None of this is easy. All of it requires sustained effort, political capital, and emotional energy that that you might not always have. There will be days when you want to give up, when things feel overwhelming, and when nothing you do seems to make a difference. These moments are not failures, and they are not unusual; they're part of the journey. The difference between leaders who succeed in chaos and those who don't isn't that the successful ones don't struggle; it's that they keep going despite the struggle.

Those leaders also know that they need to manage that struggle, that stress, and those outcomes of expending all that emotional energy. You need to find outlets outside of work: friends, family, hobbies, and pets—whatever works for you. We both write articles and books in our spare time. (Though we don't recommend this as an output.) Juan found solace in photography because film develops faster than organizations do. James continues to build productivity tools because code is honest in a way that management rarely is. And, as we've mentioned, sometimes those outlets are not enough. There's no shame in walking away from something to protect your mental and physical health.

The Practice: Making It Real

So, how do you implement all of this? How do you apply the concepts, frameworks, and principles that we've discussed to make them a reality in your specific situation? The answer is: gradually, intentionally, and with numerous adjustments along the way.

Begin by taking stock of your current situation. Use the assessments and exercises that we've provided throughout the book. You don't need to do them all at once, but pick the areas that feel most urgent or most achievable and start there. It may involve establishing basic technical principles, or maybe it's implementing a simple RFC process. It could be as simple as starting daily standups that actually work. Whatever it is, pick one thing and commit to it for at least a month.

As you implement that one thing, pay attention to how it lands with your team. Are they embracing it or resisting it? Is it solving the problem you thought it would, or creating new ones? Is it sustainable given your current constraints, or does it require constant energy to maintain? Use these observations to adjust your approach. Maybe the daily standup meeting needs to be asynchronous because of time zones, or maybe the RFC process needs to be even more lightweight. Perhaps the technical principles need to be more specific or more general.

Once that first focus area is working reasonably well, add something else. Don't take on everything at once—just one more thing. Maybe now that you have daily standups working, you can add a lightweight planning session. Maybe now that you have basic technical principles, you can start code reviews that actually reference them. Maybe now that you have an RFC process, you can start tracking cycle time to see if it's actually helping.

This incremental approach might feel slow, especially when the chaos around you is screaming for immediate solutions to everything. But it's the only approach that actually works in chaos. Trying to change everything at once just adds to the chaos. It overwhelms your team, dilutes your focus, and usually results in nothing really sticking. It's better to make one real improvement that lasts than 10 superficial changes that evaporate under pressure.

Remember, too, that you don't have to do this alone. Look for allies within your team and across the organization. Find the senior engineer who's been holding things together (or hire one to do the holding) and make them your partner in change. Connect with the product manager who shares your vision for how things could work better. Build bridges with the operations team that's dealing with the consequences of the current chaos. These relationships are force multipliers, relationships that enable you to drive change across multiple fronts simultaneously.

The Future: Beyond Survival

Throughout this book, we've focused primarily on surviving and thriving in chaos. But what about after? What happens when the chaos subsides, when stability emerges, and you can finally catch your breath? The answer is that the capabilities you've built in chaos become superpowers in stability.

Teams that have learned to make decisions with incomplete information become incredibly agile when they have good information. Engineers who have learned to ship imperfect solutions iteratively become masters at continuous delivery when they have the luxury of more time. Leaders who have learned to maintain psychological safety under extreme pressure create incredibly innovative cultures when the pressure eases.

The technical principles that guided you through chaos become the foundation of a strong engineering culture. The lightweight processes that kept you moving become the seeds of more sophisticated systems. The metrics that helped you navigate uncertainty become the basis for continuous improvement. The relationships forged in crisis become the networks that drive future success.

This is perhaps the most important message of this book: the work you do in chaos isn't just about survival; it's about building the future. Every improvement you make, every capability you build, and every relationship you strengthen is an investment that will pay dividends long after the current crisis has passed. You're not just fighting fires; you're building a fireproof foundation.

The Call: Your Turn to Lead

So, here we are, at the end of our journey together through the chaos of engineering leadership. We've shared our experiences, our frameworks, our successes, and failures. We've tried to provide you with tools and techniques, principles and practices, as well as warnings and encouragement. But now, it's your turn.

You're the one in the arena, facing your specific chaos with your particular team in your unique context. No book can tell you exactly what to do in your situation. No framework can perfectly capture the complexity you're dealing with. No principle can resolve every tension you face. Leadership in chaos is ultimately about judgment, courage, and persistence in the face of uncertainty.

But you're not starting from zero. You have your experience, your instincts, and your relationships. You now have the knowledge that we've shared and the wisdom of others who have walked this path. You have a team that, despite everything, shows up every day and tries to build something meaningful. You can make things better, even if it's just a little bit at a time.

The chaos you're facing isn't a bug in the system; it's increasingly the norm for the system. The world is moving fast, technology is becoming more complex, and organizations are more interconnected than ever. The ability to lead effectively in chaos isn't just a nice-to-have skill; it's become essential for engineering leadership at all levels.

This means that the work you're doing matters more than you might realize. You're not just keeping your team productive or your systems running. You're developing and demonstrating a critical capability that organizations desperately need. You're showing that it's possible to build great software and great teams even when everything is uncertain. You're proving that engineering excellence doesn't require perfect conditions.

Remember that leadership in chaos is not about being perfect but about being present. It's about showing up consistently, even when you don't have all the answers. It's about making decisions when waiting isn't an option. It's about protecting your team while pushing them to grow, maintaining hope when things look bleak, and upholding humility when things go well.

Remember, too, that you don't have to be a hero. Despite our occasional metaphors about storms, engineering leadership isn't about grand gestures or dramatic interventions. It's about the accumulation of small, positive actions over time and the conversations that prevent a misunderstanding. Leadership is strongest in the processes that eliminate a recurring problem, in the decisions that unblock a team, and in the feedback that helps someone grow. These small acts, repeated consistently, create the conditions for success.

Take care of yourself along the way. Leadership in chaos is exhausting, and burnout is very real. The same psychological safety you create for your team needs to extend to yourself. The same permission to fail that you give others needs to apply to your own efforts, and the same focus on sustainable pace that you preach needs to govern your own work. You can't pour from an empty cup, and your team needs you to be able to sustain this effort over the long haul.

Build your support network. Find other engineering leaders who understand what you're going through. Join communities, attend meetups, and participate in online discussions. Share your experiences and learn from others. The specific solutions might not transfer, but the patterns do—and the emotional support certainly does. Knowing that others have faced similar challenges and survived can make all the difference on the hard days.

Keep learning, but be selective about it. There's an endless stream of blog posts, books, podcasts, and conference talks about engineering leadership. Not all of it will be relevant to your situation and not all of it will be good advice. Learn to filter for what's useful, what's applicable, and what's worth your limited time and energy. Sometimes the best learning comes not from consuming more content but from reflecting on your own experiences.

Finally, maintain perspective. The chaos you're dealing with today will eventually pass (and, in fairness, it might pass because your organization doesn't survive). The crisis that seems overwhelming now will become a story that you tell future teams. The impossible challenge will become a capability that you've built. This doesn't minimize the difficulty of what you're facing, but it reminds you that it's temporary. You will get through this, your team will get through this, and you'll all be stronger for it.

The End, and the Beginning

As we close this book, we want to leave you with one final thought: engineering leadership in chaos is not a problem to be solved but a practice to be developed. You don't "fix" chaos and then move on. You develop the skills to navigate uncertainty, the judgment to make decisions with incomplete information, and the resilience to keep going when things get tough.

You've chosen a difficult path. Engineering leadership is challenging enough in stable conditions. In chaos, it can feel impossible. However, it's also incredibly rewarding. There's something pretty cool about taking a group of individuals and helping them become more than the sum of their parts. It's meaningful to create order from disorder and build something valuable despite overwhelming odds. Sometimes you can prove that chaos doesn't have to mean dysfunction, that uncertainty doesn't have to mean paralysis, and that difficulty doesn't have to mean failure.

Your team needs you—not to have all the answers, but to help them find answers; not to eliminate all uncertainty, but to help them navigate it; not to protect them from all challenges, but to help them grow through challenges. They need you to be the leader we've described throughout this book: present, consistent, thoughtful, and human.

Your organization needs you, too, whether they fully realize it or not. Who can deliver results despite imperfect conditions? Who can build great teams without unlimited resources? Who can make progress without perfect information? Who can maintain technical excellence while navigating business reality? Well, we think it's you. The skills you're developing and the capabilities you're building are exactly what most organizations need.

So, go forth and lead! Take what we've shared and make it your own. Adapt it to your context, evolve it based on your experience, and share what you learn with others. Build great software, but more importantly, build great teams. Create technical excellence, but also create human connections. Drive results, but also drive growth, human and otherwise.

This is your time. This is your challenge. This is your opportunity to make a real difference in your team, your organization, and your own career. The path won't be easy, but it can be worthwhile. And remember, you're not alone: lean on your support network for advice (and comfort) or as a place to vent.

Welcome to engineering leadership in chaos. It's messy, it's difficult, and it's often (or occasionally) thankless. However, it's also where the real work happens, where important problems get solved, and where the future is built. One small improvement at a time, one decision at a time, and one day at a time, you're making things better. And that's what leadership really means.

Keep going. Keep learning. Keep leading. The chaos needs you, and more importantly, your team needs you. You've got this, even when it doesn't feel like it. Especially when it doesn't feel like it—that's when real leadership happens.

Now put down this book, take a breath, and make something better than it was yesterday. That's the work. Chaos isn't going anywhere. But neither are you. And that's what matters.

Index

A

absence of structure (chaos symptom), 2
ADRs (architecture decision records), 200
Agile processes, 97-98
AI adoption
    case study, 224
    effect on performance metrics, 223-224
asynchronous communication
    chat and messaging, 183
    email, 184
audio communication, benefits, 182-183
automation technical principle, 155
availability, chat expectations, 184

B

backlogs
    establishing cadence, 112-113
    grooming sessions case study, 112
    management case study, 115
    managing, 111-112
bias toward action, 44
    sustainability, 45
blame culture (chaos symptom), 4
bottlenecks, identifying, 150
budgeting, 123
    aligning with goals, 136
    budget templates, creating, 137-138
    budgets, creating, 131
        historical data, 132
    competitive advantage, 129
    cost management, 124
        income side, 125
        project priorities, 126
    expense categories, 132

headcount
    business case for, 136
    case study, 134
    cost considerations, 133
    failure scenarios, 135
    growth and attrition, 134
    retention adjustments, 136
    role and seniority considerations, 135
    timing hiring, 134
hiring scenarios case study, 135
innovation and growth, 129
managing and tracking budgets, 138
organizational learning, 129
performance monitoring, 127
postmortems, 138
resource optimization, 127-128
risk mitigation, 127
stakeholder management, 128
strategic alignment, 124
    case study, 124
build-versus-buy
    case study, 139
    performance metrics, 214
    vendor management, 139-140
bureaucratic control, disadvantages, 44
burnout
    chaos symptom, 3
    compared to resistance, 39
    reducing through clarity, 88-89
business outcomes
    compared to process improvements, 47
    context diagnosis, 49
    improving focus, 48
    overfocusing examples, 50-51

overfocusing symptoms, 49-50
political considerations, 48-49
sustainable focus, 51

C

CAC (customer acquisition cost), 125
cadences
    activities, 112-113
    communication, 180
capability frameworks, 67-68
career opportunities, chaos as means for, 6-7
case studies
    AI adoption, 224
    backlog management, 112, 115
    budgeting and strategic alignment, 124
    budgeting for headcount, 134
    budgeting for hiring scenarios, 135
    build-versus-buy, 139
    code reviews, 193
    communication about priorities, 118
    context-solution matches, 43
    context-solution mismatches, 42
    continuous improvement, 204
    documentation, 178
    drift, 77
        damage caused by, 79
        focusing on one priority, 84
    ethics, 143
    Goodhart's Law, 219
    impact-versus-effort matrix, 109
    inspiring confidence, 95
    LTV (lifetime value), 126
    new responsibilities cost considerations, 162
    noniterative work, 119
    observing team functioning, 62
    outsourcing, 141
    performance metrics, 208, 210
        measuring individuals, 218
    protecting work capacity, 65
    psychological safety, 60, 202
    resource optimization, 128
    RFCs, 195
    switching roles, 56
    team dynamics, 66
    technical principles workshop, 157
    technical principles, cultural values, 176
    technical strategies, avoiding overcomplication, 159
    technical strategy

trade-offs, 169
    training and support, 170
    technical strategy granularity considerations, 163
    user stories, 103
    velocity performance metric, 216
    vendor relationships, 141
celebrations, confidence promoting, 99
change failure rate metric, 212
channels, communication
    purpose-built, 183
    support, 190
chaos
    activity compared to progress, 47
    assessing, 1
    bias toward action, 44
        sustainability, 45
    bureaucratic control
        disadvantages, 44
    business outcomes, political considerations, 48-49
    causes, 75
    challenges posed by, 7-8
    creating order from, 14
    decision making, need for, 46
    difficulty of managing, 234-235
    drift
        alignment, 85
        broken alignment, 85
        causes, 76-77
        compared to direction, 79-81
        creating alignment, 86
        damage caused by, 79
        diagnostic tool, 77-78
        indecision, 82
        lack of clear direction, 81
        lack of consensus, 82
        leadership challenges, 83-84
        multiple critical priorities, 81
        recovery cycle, 90-91
        shifting roadmaps, 82
        symptoms, 78
    evolution beyond, 233-234
    intervention approaches, 14
    need for control, 43
    opportunities, 6-7, 17
    orchestrating for success, 228
    outcome frameworks, symptoms versus causes, 47-48

political capital, 46
post-chaos stability, 236
process improvements compared to business outcomes, 47
providing direction, 229-230
psychological safety, importance of, 229
self-care challenges, 8-9
symptoms, 2-6
unclear business outcomes, 48
value of momentum, 231
working your way out of, 46
chat
    collaboration, 183-184
    response time and availability, 184
clarity
    creating, 88-89
    exercise, 90
    maintaining, 88
code reviews
    case study, 193
    cross-functional collaboration, 193-194
    guidelines, 194
    voice messages, advantages, 182
coding practices, cross-functional collaboration, 192
COGS (cost of goods sold), 125
collaboration
    ADRs, 200
    audio, benefits, 182-183
    backlog management, 111-112
    chat, 183-184
    communication
        challenges, 179
        establishing effective cadence, 180
        multimodal approach, 180
        rigid compared to creative, 177-179
        shared context, 179
    continuous improvement, 204-205
    decision making, 173
    documentation, importance of, 186-187
    email
        benefits, 184-185
        weaknesses, 184-185
    in-person interactions, 185
    leadership role, 201
    messaging, 183-184
    process changes, 98-99
    psychological safety, 191, 201-202
    RFCs, 194-200

    shared language, 191
    technical strategy development, 164
    value of, 232-233
    video calls, effective use, 181-182
    visual tools, 190
    visualization tools, 181
    visualizations, advantages, 181
command-and-control communication, 178
communication
    ambiguity, 188
    audio, benefits, 182-183
    challenges, 179
    channels, purpose-built, 183
    chat, 183-184
    creating estimates, 114
    cross-functional, requirements, 189-192
    documentation case study, 178
    email
        benefits, 184-185
        weaknesses, 184-185
    establishing effective cadence, 180
    handling disagreements, 187-188
    in-person interactions, 185
    leadership role, 201
    messaging, 183-184
    multimodal approach, 180
    poor (chaos symptom), 4
    priorities, 117-118
        case study, 118
    process changes, 98-99
    psychological safety, 201-202
    recordings as learning resources, 182
    rigid compared to creative, 177-179
    shared context, 179
    strong teams, 69
    support channels, 190
    team-building exercise, 189
    tools, 180
    transparency, 188
    understanding, getting confirmation, 187
    value of, 232-233
    video calls, effective use, 181-182
    visualizations, advantages, 181
competitive advantage, budgeting, 129
conceptual documentation, 186
consistency (technical principle), 155
context diagnosis
    business outcomes, 49

context understanding versus goal setting, 47-48

context-solution matches, 42-43
   case study, 43

context-solution mismatches, 42
   case study, 42

new employees, rapid assessments, 41-42

power and identity assessments, 40-41

context ignorance, performance metrics, 220

context-specific metrics, 213

continuous improvement, 204-205
   case study, 204

cost management, 129-130
   budgeting, 124
   fixed costs, 130
   human side, 131
   income side, 125
   project priorities, 126
   step costs, 130-131
   variable costs, 130
   vendor management
      build-versus-buy, 139-140
      relationships, 140
      selecting vendors, 140

cost of goods sold (COGS), 125

costs (see budgeting; cost management)

crisis mode, frequency of as chaos symptom, 5

cross-functional collaboration
   coding practices, 192
   coding review, 193-194
   requirements, 189-192

cross-functional connections, importance of, 23

cultural context, business outcomes, 49

cultural dimension (diagnosis context), 37

cultural lens, problem solving, 40

cultural values, technical principles, 175
   case study, 176

current technology landscape technical strategy document section, 165

customer acquisition cost (CAC), 125

D

data sources, performance metrics, 214

data-driven decision making, 203

decision lifecycle, RFCs and ADRs, 200

decision making
   collaboration, importance of, 173
   communication, 118
   data-driven, 203

frameworks for prioritization, 102
   user stories, 103-104
   user story acceptance criteria, 105
   user story impact-versus-effort matrix, 106-111
   user story limitations, 105

inclusive, 203

key inputs for prioritization, 101-102

monitoring progress, 116-117

post-decision retrospectives, 204

transparency, 203

delivery metrics, 213

demand characteristics, performance metrics, 220

deployment frequency metric, 212

development process, balancing priorities, 114

development toolchains, DORA metrics support, 214

DevOps Research and Assessment (DORA) metrics, 212

distributed teams, off-site gatherings, 185

documentation, 200
   (see also RFCs (requests for comments))
   email, 184
   importance of, 186-187
      case study, 178
   leadership investment in, 187
   maintenance, 186
   technical principle, 175
   technical strategy, contents, 164-167

documentation technical principle, 155

documenting leadership role responsibilities
   action plans, 24
   assessing planning approach, 27
   assessing team impact, 33
   diagnosing mission disconnections, 24
   key responsibilities, 21
   leadership approach, 33
   leading for outcomes, 33
   leading indicators, 24
   measurable outcomes, 24
   planning approach, 27
   process survey, 29
      issue fixing, 30
      iteration, 31
      result assessment, 30
   role descriptions, 27
   team end state, 20
   team starting point, 20

DORA (DevOps Research and Assessment)
    metrics, 212
    development toolchain support, 214
drift
    alignment, 85
        broken, 85
        creating, 86
    case study, 77
        damage from, 79
        focusing on one priority, 84
    causes, 76-77
        indecision, 82
        lack of clear direction, 81
        lack of consensus, 82
        multiple critical priorities, 81
        shifting roadmaps, 82
    compared to direction, 79-81
    damage caused by, 79
    diagnostic tool, 77-78
    leadership challenges, 83-84
    recovery cycle, 90-91
    symptoms, 78

E

email
    benefits, 184-185
    weaknesses, 184-185
emergencies, frequent (chaos symptom), 5
engineer (leadership role), 53-54
engineering leadership, core pillars, 11
engineering managers, performance metrics, 209
engineers, performance metrics, 209
estimates
    developing consistency, 113-114
    selecting methods for, 113
    using effectively, 94
ethics
    areas of responsibility, 142-144
    case study, 143
    training employees, 144
evolution technical principle, 155
expense categories, creating budgets, 132

F

finances (see budgeting)
fixed costs, 130
future state landscape technical strategy document section, 165

G

given-when-then acceptance criteria, 105
goal setting
    budgetary alignment, 136
    versus context understanding, 47-48
Goodhart's Law, 219
governance and decision-making processes
    technical strategy document section, 167
gross margin, 125
growth, budgeting, 129

H

headcount
    budgeting
        growth and attrition, 134
        timing hiring, 134
    business case for cost of, 136
    cost considerations, 133
    hiring failure scenarios, 135
    retention costs, 136
    role and seniority cost considerations, 135
hiring considerations, 20
human context, business outcomes, 49
human dimension (diagnosis context), 37
human lens, problem solving, 39

I

impact sessions, 24
impact-versus-effort matrix
    case study, 109
    exercise, 110-111
    user stories, 106-111
implementation roadmap technical strategy document section, 165
in-person interactions, collaboration, 185
information flow, multimodal approach, 180
innovation
    budgeting, 129
    effect of chaos on, 3
intervention approaches, chaos mitigation, 14
introduction and overview technical strategy document section, 164
inventories (see technical inventories)
invisible work, 220
iterative work, running in parallel with linear work, 121

L

lack of ownership (chaos symptom), 2
language (shared), collaboration, 191
lead time for changes metric, 212
leadership, 237-238
    adaptability, 52
    adapting to change, 87
        maintaining clarity, 88
    alignment, creating, 86
    breadth of knowledge, 16
    celebrations, 99
    challenges, drift, 83-84
    complexity of, 13
    cultural lens, 40
    documenting responsibilities, 17
        action plans, 24
        assessing planning approach, 27
        assessing team impact, 33
        defining leadership approach, 33
        defining measurable outcomes, 24
        defining planning approach, 27
        diagnosing mission disconnections, 24
        key responsibilities, 21
        leading for outcomes, 33
        leading indicators, 24
        process survey, 29
        process survey issue fixing, 30
        process survey iteration, 31
        process survey result assessment, 30
        rationale for, 18
        role descriptions, 27
        team end state, 20
        team starting point, 20
    focus on execution, 95
    gauging your influence, 19
    hiring considerations, 20
    importance of listening, 18
    inspiring confidence, 94-95
        case study, 95
    job descriptions, need for adaptability, 15
    leading by example, 171
    mission
        connecting to company mission, 22
        creating clarity, 22
        instilling sense of purpose, 23
        measurable objectives, 23
    mission clarity, 19
    morale, 19
    need for action, 17

new responsibilities, 16
people lens, 39
performance management, 20
performance metrics, 209
pillars, 13-15
planning
    adjustments, 26
    checkpoints, 26
    delegation, 26
    execution, 26
    partnerships, 25
    roadmaps, 25
prioritization
    changing work requirements, 100
    process methodologies, 100
    workloads, 100
process
    adjustments, 29
    avoiding complexity, 28
    team involvement, 29
product
    balancing quality and speed, 32
    defining success, 31
    delivering value, 31
    recognizing hidden wins, 32
psychological safety, 202
role in collaboration and communication, 201
roles
    case study, 56
    engineer, 53-54
    exercise, 56-58
    medic, 53
    pilot, 52-53
    rigidity, 51
    shifting between, 51
    shifting constraints, 54
    shifting skillset as senior leader, 55
    shifting skillset needed, 54
setting priorities, roadmaps, 25
struggling teams
    first steps toward functionality, 61
    removing blockers, 63-64
    resetting expectations, 62-63
    understanding team state, 61-62
system lens, 38
team dynamics, understanding, 65-66
team work capabilities, building, 67
team work capacity, protecting, 64-65

understanding team dynamics, 18
work lens, 39
leadership approaches, organizational stages, 38
leadership range, 35-36
context lenses, 38-40
diagnosing context, 37
diagnosis dimensions, 37
legal considerations for outsourcing, 142
lifetime value (see LTV)
lightweight processes, advantages, 96-98
linear work, running in parallel with iterative
work, 121
LTV (lifetime value), 125
case study, 126
lunch-and-learns, 183

M

maintainability technical principle, 155
measurements
activity compared to progress, 47
lack of meaningful (chaos symptom), 6
outcome frameworks, symptoms versus
causes, 47-48
processes, 98
medic (leadership role), 53
messaging, collaboration, 183-184
metric overload, 212
metrics, 208
(see also performance metrics)
monitoring progress, 116-117
tracking, 24
value of, 231-232
mind maps, 151-153
mission (leadership pillar)
intervention approach, 14
leadership
connecting to company mission, 22
creating clarity, 22
instilling sense of purpose, 23
measurable objectives, 23
management responsibilities, 15
modularity technical principle, 154
monitoring, metrics, and continuous improvement technical strategy document section,
167
monthly impact sessions, 24
morale, nurturing, 19
multicasting, compared to spamming, 181
multimodal communication, 180

N

noniterative work
planning, 120
prioritization, 118
case study, 119
linear workflows, 119
nonurgent communication, email, 184

O

onboarding challenges, undefined processes, 3
open source solutions, 45
openness technical principle, 155
operational dimension (diagnosis context), 37
opportunities in chaos, 6-7
organizational change, how-to overview, 9-10
organizational debt, identifying, 116
organizational learning, budgeting, 129
organizational structure, absence of, 2
outcomes (see business outcomes)
outsized responsibility (chaos symptom), 3
outsourcing
case study, 141
managing, 141-142
ownership
performance metrics, 221
step costs, 131

P

partnerships, leadership planning, 25
people (leadership pillar)
gauging your influence, 19
hiring considerations, 20
intervention approach, 14
leadership, importance of listening, 18
management responsibilities, 15
mission clarity, 19
nurturing morale, 19
performance management, 20
understanding team dynamics, 18
people, resources, and teams technical strategy
document section, 165
performance awareness technical principle, 155
performance metrics
advantages and disadvantages, 208
AI adoption considerations, 223-224
case study, 208, 210
context ignorance, 220
context-specific metrics, 213

DORA, 212
  fit to audience, 216
    team-level measurements, 217-218
  focus of, 208
  Goodhart's Law, 219
    case study, 219
  implementation strategy, 213
  intentional design, 209
  inventorying currently tracked data, 214
  measuring individuals case study, 218
  purpose, 224
  qualitative versus quantitative data, 218
  removing, 222
  selecting, 212-214
  team members, measuring individuals, 217
  unmeasured work, 220
  updating and versioning, 221-223
  user research
    crafting questions, 211-212
    deciding what to measure, 209-211
  velocity
    case study, 216
    issues with, 215
    properly tracking, 216
  visibility, 221
performance monitoring, budgeting, 127
performance, managing, 20
personality, in communication channels, 184
physical environment, collaboration quality,
  190
pilot (leadership role), 52-53
plan (leadership pillar)
  intervention approach, 14
  leadership
    adjustments, 26
    checkpoints, 26
    delegation, 26
    execution, 26
    partnerships, 25
    roadmaps, 25
    setting priorities, 25
  management responsibilities, 15
planning, lack of (chaos symptom), 5
podcasts, 183
post-decision retrospectives, 204
power assessments, context diagnosis, 40-41
priorities
  monitoring progress, 117
  transparent communication about, 117-118

case study, 118
prioritization, 25
  changing work requirements, 100
  financial considerations, 126
  frameworks for decision making, 102
    user stories, 103-104
    user story acceptance criteria, 105
    user story impact-versus-effort matrix,
      106-111
    user story limitations, 105
  impact-versus-effort matrix exercise,
    110-111
  key inputs for decision making, 101-102
  noniterative work, 118
    case study, 119
    linear workflows, 119
    planning, 120
  product development work categories, 114
  running linear and iterative work in parallel,
    121
  systems architecture diagrams, 150
  tools, 103
prioritizing
  process methodologies, 100
  workloads, 100
privacy, outsourcing considerations, 142
problem solving
  cultural lens, 40
  human lens, 39
  system lens, 38
  work lens, 39
procedural documentation, 186
process (leadership pillar)
  enabling execution, 230-231
  intervention approach, 14
  leadership
    adjustments, 29
    avoiding complexity, 28
    team involvement, 29
  management responsibilities, 15
processes
  changes, communication and collaboration,
    98-99
  focus on execution, 95
  lightweight, advantages, 96-98
  measurements, 98
  methodologies, prioritizing, 100
  rules, 96
product (leadership pillar)

intervention approach, 14
leadership
  balancing quality and speed, 32
  defining success, 31
  delivering value, 31
  recognizing hidden wins, 32
  management responsibilities, 15
product development, balancing priorities, 114
product management
  managing backlogs, 111-112
  monitoring progress, 116-117
programming, in-person interaction, 185
psychological safety
  benefits, 202
  case study, 60, 202
  collaboration, 191
  communication and collaboration, 201-202
  creating, 59-60
  importance of, 228-229

Q

quality metrics, 213

R

recordings, as learning resources, 182
recrimination (chaos symptom), 4-5
reference documentation, 186
regulations, outsourcing, 142
reliability metrics, 213
reliability technical principle, 154
remote teams, collaboration, 190
requests for comments (see RFCs)
resistance compared to burnout, 39
resource optimization
  budgeting, 127-128
  case study, 128
response time, chat expectations, 184
responsibilities
  lack of ownership, 2
  outsized (chaos symptom), 3
review sessions, backlog management, 111
RFCs (requests for comments), 194
  case study, 195
  creating, 196-200
risk management and contingency plan-
    ning technical strategy document section,
    166-167
risk mitigation, budgeting, 127
roadmaps

importance of, 25
quarterly reviews, 24
roles
  engineer, 53-54
  medic, 53
  pilot, 52-53
  rigidity, 51
  shifting between, 51
    case study, 56
    constraints, 54
    exercise, 56-58
    skillset needed, 54
    skillset needed as senior leader, 55

S

scalability technical principle, 154
scapegoating (chaos symptom), 4
security by design technical principle, 155
self care, challenges, 8-9
shared language, collaboration, 191
simplicity first technical principle, 154, 175
single points of failure, identifying, 150
spam compared to multicasting, 181
sprints, 112
stakeholder management, budgeting, 128
startups, systemic pressures, 38
step costs, 130-131
strategic vision and objectives technical strategy
    document section, 164
strategy
  lack of (chaos symptom), 5
  nonfunctional needs, 148
switches, leadership range, 35-36
  context lenses, 38-40
  diagnosing context, 37
  diagnosis dimensions, 37
synchronous communication, chat and messag-
    ing, 183
system context, business outcomes, 49
system lens, problem solving, 38
systems architecture diagrams, 149-150

T

task responsibilities, lack of ownership, 2
team culture, in communication channels, 184
team health metrics, 213
teams
  capability frameworks, 67-68
  characteristics of strong, 68, 229

communicating with clarity, 69
handling ambiguity, 68
identifying risk, 70
investing in team health, 70
managing cross-team dependencies, 69
measurable business outcomes, 71
understanding capacity limitations, 69
cross-functional relationships, 232-233
dynamics
case studies, 66
understanding, 65-66
growth considerations, 71
leadership role changes, 72-73
observing functionality case study, 62
struggling
first steps toward functionality, 61
removing blockers, 63-64
resetting expectations, 62-63
understanding team state, 61-62
work capabilities, building, 67
work capacity, protecting, 64-65
technical debt, quantifying and addressing, 115-116
technical inventories, creating, 149
technical leads, importance in planning process, 25
technical principles
characteristics, 153-154
creation process, 156-157
cultural values, 175
case study, 176
developing, 174-179
displaying, 176
documentation, 175
examples, 154-155
implementation and buy-in, 158
list of, 174
modeling, 176
overly detailed, 174
providing direction, 230
purpose, 175
simplicity first, 175
technical strategy document section, 165
updating, 176
usefulness in decision making, 194
workshop case study, 157
technical strategy, 147
avoiding overcomplication, case study, 159
build versus buy considerations, 158-160

collaborative development, 164
document
contents, 164-167
keeping relevant, 167
flexibility, 164
granularity considerations, 163
implementing, 167
communication, 170
communication considerations, 168
leading by example, 171
monitoring, 169
resource allocation, 168
risk management, 168
technology selection considerations, 168
trade-offs, 169
training and support, 170
mind maps, 151-153
new technologies, adoption considerations, 160-162
scaling, avoiding overcomplication, 158-160
technical principles
characteristics, 153-154
examples, 154-155
trade-offs case study, 169
training and support case study, 170
uses and features, 148
testing technical principle, 155
threads (communication channels), maintaining coherence, 184
time to restore service metric, 212
turnover, high (chaos symptom), 5

U

undefined processes (chaos symptom), 3
unmeasured work, performance metrics, 220
user research, performance metrics
crafting questions, 211-212
deciding what to measure, 209-211
user stories
case study, 103
epics, 105
prioritizing work, 103-104
acceptance criteria, 105
impact-versus-effort matrix, 106-111
limitations, 105
templates, 104

V

variable costs, 130

velocity performance metric
    case study, 216
    issues with, 215
    properly tracking, 216
vendors
    managing
        build-versus-buy, 139-140
        case study, 141
        relationships, 140
    selecting, 140
versions, performance metrics, 221-223
video calls
    effective use, 181-182
    tools, 182
virtual environment, collaboration quality, 190
visibility technical principle, 155
visual collaboration tools, 190
visualizations
    advantages, 181
    progressive disclosure technique, 181
    tools, 181
    visual language standardization, 181

W

whiteboarding compared to in-person interactions, 185
work capabilities, building, 67
work capacity, protecting, 64-65
    case study, 65
work context, business outcomes, 49
work lens, problem solving, 39
workflows, linear
    planning, 120
    prioritization, 119

About the Authors

Juan Pablo Buriticá knows what it takes to build and scale distributed engineering teams that thrive—whether at fast-moving startups or public companies. As he writes this book, he leads global technology at Convergint as CTO. He has built product engineering teams at Stripe and Splice, always focused on technical excellence and strong team culture. Beyond his work in the industry, he has helped bootstrap the tech community in Colombia and across Latin America through conferences, meetups, and continuous community advocacy, including founding one of the largest Spanish-speaking JavaScript collectives.

James Turnbull is CTO at Smartrr, an ecommerce startup focusing on subscriptions and postsale experiences for customers and merchants. Before Smartrr, he was SVP of Engineering at Sotheby's, ran startup advocacy at Microsoft, and was founder and CTO at Empatico, CTO at Kickstarter, VP of Engineering at Venmo, and in leadership roles at Glitch, Docker, Timber, and Puppet. He also had a long career in enterprise technology, working in banking, biotech, and ecommerce. He chaired the O'Reilly Velocity conference series, is a startup advisor and investor, and has written 11 technical books.

Colophon

The animal on the cover of *Engineering Leadership: The Hard Parts* is the emu (*Dromaius novaehollandiae*). Native to Australia, it's best known for its height and inability to fly. As the third-tallest bird in the world, it can grow up to 1.9 meters (6 feet 3 inches). Its wings are considered vestigial, but it does sometimes flap them to stabilize itself when running, at speeds of up to 48 kilometers (or 30 miles) per hour.

Its diet is omnivorous—insects, vegetation, flowers, fruits, and more. Since emus don't have teeth, they also sometimes swallow small rocks to help grind the food into smaller pieces in their gizzard. They can go for long stretches without eating or drinking.

Female emus mate multiple times in a season, laying 5 to 15 beautiful dark-emerald eggs (one every few days) before continuing on in search of other mates. The males take their parental role seriously, guarding the nest (which may house eggs from different females) and barely eating or drinking until the chicks hatch about 8 weeks later. The males then care for the young until the next breeding season.

Many of the animals on O'Reilly covers are endangered; all of them are important to the world.

The cover illustration is by José Marzan Jr., based on an antique line engraving from *Johnson's Natural History*. The series design is by Edie Freedman, Ellie Volckhausen, and Karen Montgomery. The cover fonts are Gilroy Semibold and Guardian Sans. The text font is Adobe Minion Pro; the heading font is Adobe Myriad Condensed; and the code font is Dalton Maag's Ubuntu Mono.

O'REILLY®

Learn from experts.
Become one yourself.

60,000+ titles | Live events with experts | Role-based courses
Interactive learning | Certification preparation

**Try the O'Reilly learning platform
free for 10 days.**